KANGAROO ETHICS;
PSYCHIATRY'S REQUIEM FOR PATIENT AUTONOMY: BIGOTRY REQUIRED

LIAM WYNNE

ISBN: 978-1-6847-1800-9 (sc)
ISBN: 978-1-6847-1799-6 (e)

Library of Congress Control Number: 2019914926

Lulu Publishing Services rev. date: 01/24/2020

CONTENTS

"The heresy of one age becomes the orthodoxy of the next."

— Helen Keller

AUTHOR'S NOTE

Out of respect for the real people who animate this story, their actual names are not used in this text. Their misfortune to be caught up in this drama is not an expedient for the creation of more difficulty in their lives. Instead, they are provided with fictitious surnames that are introduced in serial alphabetical order, to ease the reader in recognizing at what points in the story their roles begin. A special thanks to Michael, for his encouragement, to Graham for his long acquaintance with the English language, and to Laurel and Jim for their insight.

PREFACE

The text of this book is divided into three sections: Case notes, Ethical Analysis and Commentary. In the beginning portion, an accurate time line will allow review of the specifics of a case involving a psychiatrist and ex-patient. Many ethical cases in the literature are presented in a highly distilled version. In the author's opinion, this simplification may be useful for heuristic purposes, but may also remove qualities of the situation, arbitrarily, that give color and contrast to nuanced issues that are integrally involved. Within those truncated, undocumented and ignored qualities are often found the human values that help to define perception, spirituality and the core meaning of a life lived. Something is lost if the effort to simplify the clinical situation so removes the human element to the extent that analysis is reduced to a formulaic and perfunctory exercise.

Our Case Notes are rich in detail and tell a story that has many subtleties that would be lost if reduced to a few paragraphs. Depth is conveyed with the abundance of detail available through the physician's notes (with corroboration of the patient's detailed memory), the full story allowed to speak a narrative.

The Ethical Analysis will employ the normative perspective afforded by systematic review of Medical Indications, Quality of Life, Patient Preferences and Contextual Features. The conclusions that are reached will refute the existing ethical position currently endorsed by the American Psychiatric Association, "Once a patient, always a patient."

In the Commentary, discussion will endorse the alternative position, "Once a patient, once a patient." The implications for current psychiatric practice are extensive. The rights of psychiatric patients to retain appropriate claim to considerations of competence and autonomy are currently seriously undervalued. The sophistication of psychiatric ethical discourse would be improved by utilizing models of ethical review otherwise widely employed in medicine.

CHAPTER ONE

THE CASE NOTES

Chapter 1

> It seems mistaken, then, to say that ethical theory is not *drawn from* cases but only *applied to* cases. Rather, cases provide data for theory and are theories' testing ground as well. Cases lead us to modify and refine embryonic theoretical claims, especially by pointing to inadequacies in or limitations of theories.[1]

Day 1: Dr. Averill is offered and accepts a contract to become a full-time staff psychiatrist at an HMO by the acting director of mental health services, Dr. Barnes. In a letter confirming the authorization of the HMO regional executive committee, compensation, wages, and obligatory probationary period until full staff privileges are conferred are laid out in detail. Dr. Averill has previously been employed by this same HMO and voluntarily left employment to pursue private practice. In recognition of this previous employment, the probationary period for Dr. Averill is calculated to end—in our timeline—at day 164. The language is specific: "After that time your probationary period will end." Negotiations are cordial, and expectations for a long employer-employee relationship are mutually held.

For the previous eight months, Dr. Averill has contracted weekend coverage on the inpatient psychiatry unit where Ms. Clayton is employed as occupational therapy aide. Dr. Averill has made her casual acquaintance at the workplace and is aware, firsthand, that her notes are exemplary. She is entrusted by senior staff with routine, independent treatment planning and organization of patient activities on both the voluntary and involuntary areas of the psychiatric unit, her work a model of collaboration and integration.

Day 77: Ms. Clayton is referred to Dr. Averill at the request of her psychologist and marital therapist, Dr. Doyle. Dr. Doyle's referral requests an evaluation and consultation regarding identified sleep difficulties. Ms. Clayton's request differs; she requests an honest assessment of whether or not her energized state amounts to a clinical condition of bipolar disorder. Her fear is that she has inherited a genetic susceptibility to bipolar mood oscillations. She reports sleeping between three and five hours per night. There are absolutely no signs of cognitive impairment, no subjective experience of racing thoughts, no concentration problems, and no difficulty in communicating. Speech is of normal rate and rhythm. She is active in her craft trade and works at the inpatient psychiatry ward as an occupational therapy aide. Her notes are widely acknowledged to be the most organized and coherent notes entered in the patients' records. Her performance at work is consistently excellent, her attendance steady and predictable. She is humorous and intelligent. The symptom of reduced need for sleep stands alone. Criteria for the threshold of the diagnosis of bipolar disorder are not reached, the diagnosis not rendered as a current concern.

Dr. Averill concludes that Ms. Clayton is minimally energized over her baseline but in no manner impaired. Sleep hygiene suggestions are reviewed along with a suggestion to consider the PRN use of an antihistamine, Benadryl, should the sleeplessness worsen or begin to disturb daytime function. No appointment is made for follow-up. Return visits are left to the discretion of Ms. Clayton if her functioning begins to resemble the dimension of mood difficulty that has been reviewed in the patient education material and discussed in depth.

Day 107: Ms. Clayton returns to Dr. Averill to establish a contract for short-term psychotherapy. She details her expectation of specifically using the opportunity to grieve the loss of her mother six months before. She remains in a marital therapy continuously with long-term psychologist

Dr. Doyle and her husband, Mr. Eason. It is agreed that all references to marital issues will be the province of the marital therapy; if raised, the issues will not be commented on except to refer them back to the ongoing therapy. Dr. Averill is informed by Ms. Clayton of a continuing difficulty with a delayed sleep onset at essentially the same level previously reported. The symptom still exists in isolation, unaccompanied by any other criteria from the *Diagnostic and Statistical Manual of Mental Disorders* that would indicate the accrual of difficulties amounting to a diagnosable bipolar disorder. Dr. Averill prescribes a limited supply of Dalmane, a sleeping medicine to be used for inducing sleep. The expectation at the outset of therapy is for a time-limited therapy of between eight and twelve sessions. A tentative schedule for therapy is discussed and agreed upon.

Day 115: Ms. Clayton details her memory of a previous major depression. Her memory, recall, and organization are precise. It is apparent by her comments that she has done a considerable amount of introspection about and processing of her mother's death, sibling relationships, and unsettled feelings concern her parents' separation when she was a youth. It is clear she has an agenda to work in the therapy, is an independent person, and requires little encouragement to proceed.

Ms. Clayton mentions material being directly worked through in her ongoing marital therapy, that Mr. Eason, her husband, is interpreting the accumulating distance in the marriage as a measure of mental illness. She laments that her husband does not understand the real reasons for the distance. She has been sadly coming to the conclusion that Mr. Eason insistently refers to her through the language of pathologizing her behavior. She intuits that her desire to continue in these circumstances has been lessening consistently over time, which sadly indicates the marriage may be over. It is expected this is clearly an area of concern that will be discussed together with her husband in marital therapy. No other comment is made.

Day 117: Mr. Eason phones Dr. Averill to inform him of the changes that he is witnessing in his wife's behavior. He is convinced that she is indeed manic. He relates that this appreciation is because of her sleep decrease, her distance from him, and her sagging performance at work. Mr. Eason infers that his wife is less than rigorously honest about the depth of her mental illness and that he has seen her do so poorly in the past that he hopes a prescription of lithium will be made in the next office visit to preclude further ravages of his wife's mental disorder.

On this day, Dr. Doyle calls and identifies the issue of the ongoing therapy and difficulties in the relationship of Mr. Eason and Ms. Clayton. At this moment, no release of information, signed by Ms. Clayton for communication with Dr. Doyle, exists. Dr. Averill informs Dr. Doyle of this fact. He states that until he has such a release, he is constrained to a position of listening, willing to continue the discussion if Dr. Doyle has information that she thinks will be helpful. Dr. Doyle for a moment is incredulous regarding Dr. Averill's hesitance to speak about details of Ms. Clayton's therapy. Dr. Doyle relates that the ongoing pressures in the marriage are such that the use of lithium may be wise, not that a definable mood disorder currently exists, but that its prescription may avert such an event.

At the time, Ms. Clayton simultaneously maintains a self-owned craft business, teaches crafts at a local craft shop, and works at the psychiatric inpatient unit. From senior staff at the unit, it is clear that her work is, even at that moment, of exceptional merit. Often, the most precise, informative, and intuitive comments in the entire chart are consistently found in her notes, true also at this time. Ms. Clayton misses no work and is seen as consistently skillful in the coordination of group activities with the most severely disturbed psychiatric inpatients, often entrusted to her care for several hours per day.

Ms. Clayton is seen this day at her request. A detailed discussion of lithium occurs for the purpose of allowing the process of informed consent. As lithium has historically been helpful in the past in a period of depression and as a preventative for the possible likelihood of recurrence, Ms. Clayton agrees to take it as a precaution. She is informed of the phone calls from Dr. Doyle and her husband. She is not happy with her husband's intrusion and continued emphasis on the nature of her supposed mental illness. She hears of Dr. Averill's reluctance to discuss details of therapy concerning her mother's death with Dr. Doyle and is appreciative of the boundary.

With boundary issues involved in the nature of so many phone calls about her, without her knowledge or acquiescence, she does not move to sign a release for Dr. Doyle. Instead, she states her desire, more characteristic for her in her own estimation, to tell Dr. Doyle directly about the status of her separate therapy.

Day 121: Ms. Clayton returns for her regularly scheduled therapy appointment. She informs Dr. Averill that she initiated a separation from her husband and that her sleep improved on that same day. The wisdom of this decision has not been discussed with Dr. Averill directly or indirectly. Neither is Dr. Averill asked to comment on the advisability or inadvisability of this action in this office visit. The timing of the sleep improvement from the date of initiating lithium is one day. Lithium requires four to five days to achieve stable concentration in the blood. It is also known to have an unfortunate delay in its onset of action beyond that time frame. It is clear that the improvement in this sleep symptom occurs so rapidly that lithium cannot be responsible.

Ms. Clayton continues her pace of bringing up details of pertaining to her mother's death by cancer. She maintains her ability to focus on the issues that promoted her desire for this short-term therapy.

Day 127: Ms. Clayton returns for a scheduled appointment. Her mental status is consistently clear of any indication of mood disorder or cognitive impairment. The content of her therapy on this date concerns her spiritual growth and how it relates to her understanding of the significance of her mother's loss. She is relating issues of acceptance, the beginning of a recognition that she must continue in her mother's absence. Her clinical presentation is without any evidence of mood pathology.

Day 130: Ms. Clayton returns for a scheduled appointment, the schedule constructed at the onset of therapy according to Ms. Clayton's stated desire to do the work without delay. The overall appreciation that the therapy has naturally come to an end point is raised by Dr. Averill. Ms. Clayton agrees and asks to do the work of closing formally in the next appointment.

Day 134: Ms. Clayton describes the therapy she has worked in this last month as satisfying her need to put the loss of her mother in perspective. She feels she was removed from this work, at the time of her mother's loss, due to the presence of a severe major depression. She believes she was restrained subsequently, as that depressive episode ended, by the realization that her marriage was dissolving. She believes she has concluded her therapy. She has some continuing misgivings about the need for continuing the Lithium that has been initiated.

Dr. Averill states his reasons why a formal termination to the therapy is warranted. First, Dr. Averill has admitting privileges to the same psychiatric inpatient ward where Ms. Clayton works. He is concerned about an ongoing doctor-patient relationship with a coworker. As the therapy has unfolded, this realization has moved Dr. Averill to consider the probity of clearly ending the doctor-patient relationship.

Second, Dr. Averill believes that Ms. Clayton's need for Lithium is not clear. Her improvement, so closely aligned to improvements in her psychosocial stressors, is not evidence of Lithium's clinical effect. Lithium is not without its potentially serious side effects on kidney and thyroid. Dr. Averill believes it is possible that she is on Lithium without sufficient cause. He moves to refer her to a second opinion for ongoing psychiatric care, to review the prescription of Lithium and take over the clinical coordination of medical care. He refers her to an acknowledged expert in psychopharmacology.

And third, Dr. Averill notes that a long-standing venue for therapeutic concerns already exists with Dr. Doyle. Even though separated for several weeks, Ms. Clayton is planning to continue in this therapy and explain, if her husband will allow, how she has decided that it is clear to her that he has little regard for her, attributes her actions to a persistently held Zeitgeist of "mental illness" and that the respect necessary for a marriage is gone. Dr. Averill notes that by bringing her therapy back to Dr. Doyle, any disruptions in what has been a very fruitful relationship with Dr. Doyle as a therapist, owing to splitting out projects of therapy to other providers, will be avoided.

Ms. Clayton voices an understanding and acceptance of this formal termination.

Day148: Ms. Clayton calls Dr. Averill with the offer of a gift in recognition of the respect shown and the value gained throughout the therapy. Dr. Averill accepts the invitation, believing Ms. Clayton to be independent of character.

It is at this moment in time that in separate locations and separate introspections, both Ms. Clayton and Dr. Averill appreciate what they understand as an unfolding, spiritually based awareness. They separately consider, for the first time, that they have discovered the person they were placed on earth to live with. Both believing themselves to be most closely Buddhist, they believe they are drawn to each other by their most sincere spiritual consciousness. Their awakening to this realization promotes contemplative journal entries and the beginning of a severe questioning of the motivations and unconscious mechanisms that may be distorting this event, true to mid twentieth century psychiatry's understanding of transference and countertransference.

Day 149--153: A wealth of written correspondence of considerable scope and intent now begins to be exchanged by Dr. Averill and Ms. Clayton. Dr. Averill addresses the dangers in the situation cogently. He describes the issues of transference and countertransference in detail. For this, he draws on a background of a residency training program unusually rich in psychotherapy education; more than five hundred hours of didactic training, and five hundred hours of direct supervision with the area's faculty, most of whom are analytically trained. Over the course of five years, Dr. Averill has trained with eighteen mentoring therapists, including the president of the analytic community. He was acknowledged in training as a solid resident with regard to psychotherapy. His family therapy mentors considered him, at the end, among the few people to whom they would refer patients.

Dr. Averill asks very hard questions about the motivations of meeting, aware of and communicating the more normal processes of waiting for sufficient time to pass. He communicates the current American Psychiatric

Association warning, to the effect that relationships with former patients are almost always unethical. He communicates his fear that from the outside, any contact may be construed as manipulative and deceitful.

Ms. Clayton voices a clear understanding of the principles of transference and counter transference. She believes these concepts to be wholly unsatisfactory in explaining the understanding she more closely attributes to a spiritual reality. She is glad that cautions exist to thwart the manipulation of women, but she is 100 percent clear she is not being manipulated by an unconscious series of defense mechanisms. She experiences herself in this moment to be acting in accord with her highest principles, aware that her decisions are her own. She is clear for herself that her relationship with Dr. Averill has never included manipulation or deceit. She asks to be her own judge of the saliency of the constructs of transference and countertransference. She is sorry for the timing of this awareness, but feels she would be sorry not to pay attention to the single most clear, unprecedented spiritual moment of her life; a recognition that she has met a person that, in as much as she is able to understand, she is intended to be related to in this lifetime. She relates this in a way that makes clear the lack of superficiality of the statements.

Dr. Averill has been aware during the process of therapy that there exists in the character of Ms. Clayton an honest, independent person. He finds in her the type of person who is fundamentally healthy.

At this moment, after the formal termination of therapy and after referral to other care providers, Averill is himself surprised to find separately an entirely consistent understanding of events, similar to those Ms. Clayton has reached. Despite the cautions of his profession, he finds it entirely likely that the understanding of the various manifestations of transference and countertransference cannot completely account for what he experiences.

The two parties, then, are clear that in the most honest reaches of their hearts they can find no other explanation more accurate than to say that they are in the process of realizing a relationship exists between them that is not illusory, not superficial, not conveyed particularly well or clearly through the language of psychiatry. It is real, in their experience as deep as their innermost spirituality, and a blessing. It is occurring with poor timing, nevertheless.

Day 152: Averill and Ms. Clayton become directly involved romantically.

Day156: Dr. Averill meets with Mr. Flannery, the supervisor of the HMO mental health service where he is employed. A concise and honest discussion of the relationship and its evolution results. Mr. Flannery is appreciative of the honesty of the sentiments and is supportive, if not cautiously guarded about the reception this news will have among the mental health staff. It is Mr. Flannery's decision not to share this information at present and to await developments.

Day 159: After considerable thought, Ms. Clayton decides to go directly to her separated husband. They have lived separately for over a month. She intends to tell him the whole story so that he will be able to hear from her directly that there was no exploitation. She is hopeful she will be afforded the chance to explain the real circumstances of her awareness. She truly believes that her husband will be able to discern the honesty and clarity of her position and, that by being respected in the act of a direct conversation, he will be less prone to defensiveness. She is, in fact, reviled. Mr. Eason tells her that she has been duped, that she has been brainwashed, that she is patently lying and that he suspected she was involved with Dr. Averill during the course of therapy.

Day 160: With Ms. Clayton in attendance, Averill calls Mr. Eason to apologize for the timing of events and to describe a sincere appreciation of the attraction for Ms. Clayton as emanating from a spiritual place in his heart. Mr. Eason is very angry. He warns Dr. Averill that things will go well for a while, then Dr. Averill will come to appreciate that he has married a very sick person.

Day 162: Mr. Eason calls Dr. Averill to inform him that he will get the doctor fired and have his license taken away. Dr. Averill calls on this day to request information about his upcoming change from probationary to full time status in the HMO. This status is due to change on Day 164, per contract.

Day 163: Dr. Averill is informed that the regional medical director, Dr. Geddes, instructed the acting Medical Director of Mental Health, Dr. Heath, to contact Dr. Averill with a disqualifying statement; that whoever worded Dr. Averill's contract with an abbreviated probationary period did so without the proper input from the advisory administrative office. Dr. Averill is informed that the contract is not considered valid and is henceforth rewritten to include a one-year period of probation. Dr. Heath is extremely apologetic and admits he was seriously remiss to create such an expectation. He is clear that he has been severely chastised by Dr. Geddes.

Dr. Averill knows that Mr. Eason will shortly be calling to inflict as much damage as possible. He decides in the moment to allow that Dr. Heath made a human error and not to pursue contractual litigation. No discussion of the presence of any complaint against Dr. Averill is mentioned.

Day 164: Dr. Averill is phoned by Dr. Heath to inform him that he will need to appear before an Ad Hoc Quality Assurance Committee. Dr. Heath offers the spontaneous comment that Dr. Geddes moved to continue Dr. Averill's probationary period due to the wording of the contract, not due to any complaint already received at the HMO.

Day 165: The Ad Hoc Quality Assurance Committee is convened in a neighboring city. Dr. Heath; the psychologist Chief of the HMO Mental Health Service, Dr. Ireland; a psychiatric colleague, Dr. Joyce, Mr. Flannery and the Chief of the Medical Staff for the HMO, Dr. Keating, are in attendance for the meeting with Dr. Averill. Dr. Averill is encouraged to consider at the outset that he might want a lawyer present. He states that he has nothing to hide. The committee covers the rules of confidentiality to clarify their understanding that the contents of their discussion will be "non discoverable." Dr. Averill is advised not to discuss the meeting with anyone.

Dr. Averill details his understanding of the relationship in full. His comments are candid. He makes clear that he did not envision any relationship with Ms. Clayton during the therapy, he formally terminated the therapy and at the time there were absolutely no plans to contact Ms. Clayton. He shares his understanding of the spiritual underpinnings of the relationship as it subsequently developed.

The committee discusses what would be in the best interest of the HMO. Dr. Heath states that he wishes he could just say the situation has been reviewed and chastise Dr. Averill privately, saying, "Don't do that again," and then move on. Dr. Ireland adds that this is a difficult situation because Dr. Averill is so personable, liked and talented. He believes many of the mental health staff will be hurt and angered by this event. He states that Dr. Averill would be well advised to volunteer a statement to the Medical Board.

Dr. Averill agrees to submit such a statement and asks to be the first to do so, if such a document is in order. It is agreed that he will be permitted to be the first representative from the HMO to submit any document to the Medical Board. Dr. Averill then volunteers to have his work reviewed regularly in supervision. Dr. Heath agrees this would be a good idea and tailors it to include the use of an outside consultant in order to confer added credibility, should public scrutiny via television or newspaper occur. The committee recommends reaffirming the issue of one year of probation, now seen as further indication of the caution the HMO is exercising in its duty to protect the public.

Day 172: Dr. Averill is given four hours' notice to cancel his appointments and appear in a neighboring city for a hastily convened meeting. Here, the HMO legal counsel, Mr. Locke, Mr. Flannery and Dr. Heath are in attendance. Discussion is held about assigning monies from the Medical staff budget for legal representation. Suggestions are offered about names of competent lawyers who have previously defended HMO physicians.

Confirmation occurs concerning the suggestions issuing from the QA Committee meeting of Day 165. The length of supervision will be detailed as one year; the content will specifically deal with transference and countertransference. The supervisor will be allowed to talk to the HMO about Dr. Averill's case discussions and, if necessary, give recommendations.

Dr. Averill is again advised to submit a statement to the Medical Board. Those present acknowledge that Dr. Averill will require time to meet with counsel to prepare a statement. It is confirmed that this situation will be handled as confidential. Dr. Heath states that he will inform Mental Health staff that the matter has been thoroughly looked into and dealt with, as far as the HMO is concerned. Dr. Averill is informed that Dr. Geddes, Regional Medical Director, has been informed of the recommendations and stands behind the approach.

Day 173: Ms. Clayton sees the psychiatrist to whom she has been referred, Dr. Marchant. At issue is the advisability of Lithium and a review of the opinion that Ms. Clayton does not currently manifest signs of a mental illness. Ms. Clayton tells Dr. Marchant that she believes the relationship with her husband had drifted inexorably from its original frame. She is aware her husband blames her illness for everything that has transpired. She asserts that she feels healthy, does not need to be taken care of, and that this issue has thrown into great relief the lack of connection in the marriage. This fact is primarily responsible and at the root of the separation. Ms. Clayton notes that the couple has a meeting with their marital therapist that day and is quite clear they will file for divorce. Recorded in Dr. Marchant's medical record of that day is the following entry:

We discussed in some detail this first meeting, her current relationship with [Dr. Averill]. She indicates they are very fond of each other, that they did not start to see each other until she had formally terminated treatment with him and that what is happening in her marriage was occurring/ would have occurred even without involvement in a relationship with [Dr. Averill]. She believes her emotional stability is what led to her current marital situation, that her husband can't seem to tolerate her being stable like she is now. We discussed the possible effect on [Dr. Averill's] professional life-she did indicate that her husband had complained to the HMO about him. She doesn't see that she has done anything unethical, realizes the impact on his career, potentially. She expressed they now are very much in love, and she hopes to marry him by next summer-I expressed my opinion that she needs to end this marriage first, deal with those feelings and take some time getting to know each other prior to entering a new marriage … she concurs.

On this date Dr. Marchant writes to the State Department of Health professional licensing services on behalf of Ms. Clay, who is seeking accreditation as a mental health counselor:

"In reviewing her history with her, the alteration in her mood when she is ill has not affected her work performance in any manner whatsoever … Since her illness is in remission and she has good insight about her own mood changes and behavior, neither her illness nor her treatment should in any way have an impact on her working with clients."

A psychiatrist who treated Ms. Clayton 11 months prior, Dr. Nichol, submits a simultaneous letter to the licensing division on her behalf:

"[Ms. Clayton] is very well informed about Bipolar disorder, and I am confident should she experience any recurrence of symptoms she will again start treatment. I believe she has the potential to be an excellent counselor."

Throughout this same period of time (for at least a week prior and the ensuing month and a half) Mr. Eason begins a telephone campaign to discredit Dr. Averill and ruin his career. For this he enlists a long-time friend of Ms. Clayton and therapist within the same HMO system, Mr. Ott. Together they attempt to bring pressure on administrative determinations. Calls are placed to the state medical board investigative office, the HMO administrators, members of the HMO mental health staff and the American Psychiatric Association.

The site of Mr. Ott's clinic is stirred by Mr. Ott's call for action. Because of Dr. Heath's decision to ensure confidentiality, no information proceeds from the administration to the staff. The staff first hear of events from the perspective of Mr. Eason and Mr. Ott. They are encouraged to believe a perfidious act has occurred.

Day 179: Dr. Geddes, the Regional Medical Director, is on vacation and misses the Administrative Work Group meeting. His support for the working decisions of the Ad Hoc Quality Assurance Committee (continuing probation, securing supervision, etc.) is not transmitted to the administration. They have already been appraised, by telephone campaign, concerning the nature of events. They hear nothing of the discussion held with Dr. Averill. No alternative information reaches them.

During this time, on the advice of counsel, the HMO submits to the Medical Board their own separate report of the possibility of harm to a patient. Although they have promised to wait until Dr. Averill can submit a report, they do not. They do not inform Dr. Averill of this action or reasons for the change of heart.

Day 183: After meeting with counsel, Dr. Averill submits a comprehensive detailing of events to the Board. Included in that document is a statement from Ms. Clayton:

"Throughout the therapy I held the utmost respect for [Dr. Averill] as a clinician and felt he approached the therapy with integrity. At no time did I feel the behavior on his part was, in any way, seductive. There was observed, in fact, a significant physical distance between us. The professional relationship which existed between [him] and I terminated on (Day 134) with [Dr. Averill] referring me back to [Dr. Doyle] for such continuing treatment as she may deem appropriate.

Currently, [Dr. Averill] and I are seeing each other. This personal relationship which we have committed to is one built upon mutual trust and was entered into by us as discerning adults. We feel strongly that the relationship exists as a separate entity from any therapy, and that it developed upon its own merits. The bond is so healthy and reciprocal that we hope to marry in the summer of [next year]. It is of grave concern to me that anyone would in any fashion contend that I have somehow been injured as a result of this relationship. I am pleased to say that I am uninjured in any way, and that I entered into this personal relationship willingly and with conviction. I am committed to a future which holds promise of equal partnership in our future marriage."

Day 185: Mr. Flannery forwards a suggestion from the clinical staff at the mental health service that they require Dr. Averill to appear before them for a review of the situation. Mr. Flannery opens the meeting but gives no direction or format for the meeting. For the course of an hour he is silent.

Dr. Averill, having discussed the propriety of disclosure with Ms. Clayton, opens by reading her statement to the Board and encouraging a discussion about autonomy. Ms. Clayton's comments are openly scoffed at by members of the clinical staff. The staff begins to voice their repulsion for the thoroughly reprehensible act they assume has transpired. They ask no factual questions. They do not ask a single question about the ethical considerations that have gone on in the minds of Dr. Averill and Ms. Clayton. A psychologist at the clinic, Dr. Preston, observes that a revision in the American Psychiatric Association's statement of medical ethics in such circumstances is certainly overdue, given the nature of professional misconduct. (Their statement that relationship with former patients *are almost always* unethical, is not correctly quoted in this gathering.) Implied, but never directly queried, is the conjecture that seduction and abuse have occurred.

In a crescendo of emotion that proceeds as a wave around the room, Dr. Preston intones, "When did you conclude that the rules didn't apply to you?"

Dr. Averill's replies that there are simply more principles and rules at stake than the one being used as a reference point. This is lost in the gather din of disapproval. Comments emanate from various members of the staff: "So what did you expect?", "What I want to know is why is this man still working here?", "Why are male physicians protected like this?", Why isn't [the HMO] going to do something about this?". Several members complain bitterly that if something is not done, they might not work there any longer. Dr. Averill makes no attempt to appease the crowd. He admits that his timing was abysmal and his judgment certainly not circumspect, but maintains that Ms. Clayton and he are aware of other principles, have acted according to rigorous honesty, and their behavior is without manipulative action or intent.

Toward the end of the hour Mr. Flannery offers one sentence, to the effect that he doesn't see why Dr. Averill doesn't just stop the relationship. From Dr. Preston, "We will be judged by others on the outside by whom we allow to work here."

Following the meeting several mental health staff write to the administration to ask that Dr. Averill be fired. A message is left on his home answering machine from a person he has never met, a person who has not spoken to either Ms. Clayton or Dr. Averill, an employee of the HMO, that she will be very happy to try to promote with all deliberate energy the firing of Dr. Averill from his position. She adds that people of his ilk should never be allowed to be employed by that agency. She does not leave her name.

Day 190: Dr. Heath arranges for an individual conversation with Dr. Averill in the interim, until the outside analyst can be contacted to arrange supervision as the Ad Hoc Quality Assurance Committee has agreed. Dr. Heath wastes no time in asking, in a singularly brusque manner, exactly what goes through Dr. Averill's head when he sees a woman in therapy.

Dr. Averill gives a thoughtful reply, informing his superior that he has graduated from a residency training program where he had an unusually enriched program of psychoanalytically oriented, psychodynamic psychotherapy supervision. He was seen as one of the program's better therapists. Dr. Averill states that he has a normal compliment of emotions including attractions, that he does not act on. Further, he has been a psychotherapy supervisor for two years. He coordinated the psychiatry teaching in a Family Practice residency training program for three years. His boundaries have been intact. There have been no similar occurrences or complaints.

Dr. Heath asks whether Dr. Averill intends to stay with the HMO, stating he doesn't want to go out on a limb for him without reason. Dr. Averill replies that he does not want his legacy to the HMO to be one of derision. If Averill hears he is going to be supported he will try to ride out the current event until reason can prevail and people can review the situation with cooler heads.

The impression is conveyed to Dr. Averill that he is being disloyal to entertain the idea of looking into his options of employment elsewhere. Dr. Averill responds that he does not appreciate being placed in the position of needing to apologize for correctly reading the peril in the situation, believing it is indeed sensible to maintain an awareness of the possibilities, should it come down to asking for his resignation.

Day 200: The HMO Regional Administrative Work Group convenes, this time with Dr. Geddes and representatives of the Ad Hoc Quality Assurance Committee for Mental Health. Also, present are members of medical staff, and representatives from the HMO legal counsel. In a closed session they pronounce the process of their meeting to be one of peer review and so the records are "non discoverable." They insist that secrecy in this meeting's content is important. (Dr. Averill, not present, is later informed through a member in attendance.)

Several attendees have been directly phoned by the telephone campaign of Mr. Eason. They have heard what they consider to be a black and white transgression of professional ethics and obvious harm to both the patient and corporation. They have struggled within the community to preserve the dignity of their corporation and are still remembered by some of the populace for a situation seven years prior when an OB-GYN practitioner had seven simultaneous lawsuits for malpractice, during which time the medical staff took no action. Discussion proceeds with an eye toward the obvious cost to the corporation of maintaining employment of a physician who will bring ridicule. The discussion is negative and very bitter. It is clear that minds were made up before the meeting opened. The suggestions of the Ad Hoc Quality Assurance Committee are swept aside. In the atmosphere of the meeting, no one steps forward to offer the alternatives that are already in motion or defend their logic.

Day 201: Dr. Averill is intercepted at lunch and told to drive immediately to a neighboring city to meet with Dr. Heath. Mr. Flannery from the clinic is there. Dr. Heath advises Dr. Averill that it would be a very effective strategy to take a leave of absence in order to evidence that Dr. Averill has the capacity for introspection. He is informed that the atmosphere does not look good. Mr. Flannery states that he believes the clinic is improving noticeably in regard to Dr. Averill, and that the situation looks quite workable and can be sustained. Acting on advice, Dr. Averill puts in a vacation request.

Day 202: A colleague of Dr. Averill, Dr. Joyce, phones Dr. Heath to take exception to the idea that a vacation/leave of absence will improve matters. He believes that Dr. Averill is showing that by continuing he gives evidence of his continuous value and character. Dr. Heath denies that he ever suggested a vacation/leave of absence. Mr. Flannery, in attendance at the original meeting, later confirms that Dr. Heath did suggest that very action.

Day 205: A psychologist in another state, Dr. Rolleston, writes to Dr. Averill, unaware of the transpiring events. He offers a job opportunity, should Dr. Averill be interested.

Day 212: The first scheduled appointment with the analytic supervisor is kept, according to the plan evolved by the Ad Hoc Quality Assurance Committee. It is a general exploratory and contractual session.

Day 213: The combined staffs of the mental health service hold a second meeting with Dr. Averill present. It is far less harsh and rigid. There are no signs within the room that anyone is still contemplating quitting, and more discussion occurs about the complicated nature of ethics in the issues of recent events.

Ms. Clayton has her next follow-up with Dr. Marchant, who wrote:

"Discussed phone call I received from [Dr. Heath] about two weeks ago (Day 199). [Ms. Clayton's] husband called to say he thought [she] was hypomanic. She is absolutely free of any hypomanic symptoms, is euthymic, has been since I've known her. Apparently [Mr. Eason] is so outraged by what has happened that he is telling everyone that she is manicky. Many friends have pulled away from her, very painful for her. She and [Dr. Averill] are getting along well. All in all, she is handling the incredible stress well."

Day 214: The Regional Administrative Work Group meets again, this time with three of the representatives of the original Ad Hoc Quality Assurance Committee, representatives of the medical staff and HMO legal counsel. The findings and suggestions of the QA Committee are shared, then resolutely disparaged and dismissed. One of the participants in the meeting makes a statement that the HMO has public relations problems throughout the area, spreading to another state. The question of immediately firing Dr. Averill is raised and rushed to completion. There is discussion to the effect that if any legal issue is raised later by Ms. Clayton, the HMO will have appeared to have acted swiftly to censure Dr. Averill, thus limiting the HMO's liability profile. No representative from the QA Committee or Administrative Work Group makes any contact with Ms. Clayton. Her statement has been made available to the QA Committee, but it is unclear whether, in the acrimonious atmosphere, it is brought to anyone's attention. Not a single person at the HMO calls to inquire after the welfare of Ms. Clayton.

At 1:00 p.m., Dr. Averill is instructed to drop everything and appear before Dr. Heath. Dr. Averill, before discussions begin, tenders a resignation. It is not accepted. Dr. Averill is informed flatly that he has been terminated, his privileges suspended, given three months' severance.

Day 216: Dr. Averill returns the call of Dr. Rolleston, who had contacted him previously from another state with the offer of working in his office. Dr. Averill candidly explains the entire process to this point. The job offer is continued and plans commence for Ms. Clayton and Dr. Averill to move in three weeks.

Day 218: Christmas.

Day 226: In the office of Professional Affairs at the hospital where Dr. Averill holds inpatient admitting privileges (not the HMO), he returns to close out his privileges formally. By chance he reads a letter from the

Director of the Hospital Psychiatry, Dr. Seahill, to hospital administration, that Dr. Averill has been let go from his position at the HMO for "sleeping with a patient during therapy." Already stunned by events, Dr. Averill leaves without a copy.

Day 246: Averill and Clayton relocate to another state. In the process of applying for staff privileges at a local psychiatric hospital, Dr. Averill informs both the Executive Director, Mr. Taine, and Medical Director, Dr. Upton, of the entire set of circumstances that follow him from his previous employment. Dr. Upton commiserates with Averill, he too is married to a woman he first met in the role of doctor and patient. They have been happily married for nearly two decades.

Day 363: The investigator for the Medical Evaluation Board, Ms. Vaile, sends a letter to Dr. Averill's lawyer. In this letter, the nature of the investigation is described:

"The focus of the complaint is alleged sexual contact with a female patient while still being her therapist."

This is the last contact the Board initiates for 683 days.

Day 368: Dr. Averill responds to Ms. Vaile:

For the record, I had no sexual contact with (Ms. Clayton) while I was her therapist. There was no allure, no promise of a relationship and no ending of therapy in order to initiate such a relationship. Termination of therapy, with referrals for follow-up, was completed. (Ms. Clayton) and I continued to experience a very healthy, mature and equal relationship. We continue to plan our marriage as (Ms. Clayton) stated in her submittal.

Day 369: Ms. Clayton writes independently to the Board:

I encourage you to refer to my (previous) statement to the Medical Board. The content remains unchanged, and I continue to warmly anticipate marriage to (Dr. Averill) in several months. I can say in all honesty and with integrity that allegations regarding sexual contact during therapy are entirely untrue. Please accept my statement with complete respect for this investigation.

Days 275-675: Ms. Vaile, after discussions with Mr. Eason, continues in the process of gathering information for the Board. No indication of the status of the investigation is shared with Dr. Averill, Ms. Clayton or counsel. Averill's counsel informs him that there is no statute of limitations regarding how long the Board has to deliberate before making a determination. Neither are they required to inform the parties whether the investigation continues or is moved to inactive status. Ms. Vaile does not talk directly to either Dr. Averill or Ms. Clayton.

- The Board interviews the ex-husband (divorce finalized).
- The Board interviews Dr. Averill's therapist, Dr. Weybourne. She disinclines the Board from regarding Dr. Averill as irresponsible, marauding or character impaired.
- The Board interviews Mr. Ott, friend of Mr. Eason.
- The Board secures the confidential notes of the HMO Administrative Work Group from Days 200 and 214. Previously, these notes and proceedings had been deemed "non discoverable."
- The Board interviews Dr. Marchant, who received care on transfer from Averill. He shares his opinion that there has been no abuse.
- The Board interviews Dr. Doyle, Ms. Clayton's concurrently treating psychologist, who had knowledge first hand of this situation as it transpired. Dr. Doyle has held a position with the corresponding Board for Psychology. Dr. Doyle tells Ms. Vaile that she does not believe that Ms. Clayton was in any way a victim, nor was she abused. She also surmises that Dr. Averill, in not being comfortable sharing information with her without a release of information, was covering for some type of relationship. The ability to discern this, she attributes to her intuition; she believes Dr. Averill is lying. In later conversations with Ms. Clayton, Dr. Doyle divulges that the process of the interview with Ms. Vaile is unnerving. It is very apparent to Dr. Doyle, an experienced clinician, that Ms. Vaile is after any and all information that can be used in a negative characterization of Dr. Averill. It is evidently clear that the

conversation continuously attempts to elicit, expectantly, negative elements of the case, dismissing or ignoring anything not fitting with that precept.

Day 631: Averill and Clayton are married in the attendance of friends and family. Ms. Clayton's closest relative is there for the occasion. He is most supportive.

Day 649: Counsel for Dr. Averill writes to the Board to inform them of the marriage, requesting a letter of final disposition:

"In fairness to my client, I do not think the matter should continue on indefinitely."

Day 670: Dr. Averill writes to the Administrative Work Group at the HMO. He wonders if, with cooler heads, they can review the statement of <u>Principles of Medical Ethics with Annotations Especially Applicable to Psychiatry</u>, approved December 1988 (at the time, current):

> "Sexual involvement with one's former patients generally exploits emotions deriving from treatment and therefore *almost always is unethical.*" (Emphasis added)

Dr. Averill adds:

> "I heartily concur with the letter and the spirit of what is written in that statement. The statement is not designed to be used as conclusive evidence against the possibility that positive relationships can occur. If that was the intent of the APA, then this specific principle would be cast in absolute terms. The APA does not maintain that it is impossible that an ethical relationship can exist between therapist and ex-client."

Day 714: The Medical Director of the HMO replies:

> "The Quality Assurance Committee did not have, nor did it purport to have the authority to make a final decision with regard to your privileges and contract. Our records indicate that on at least two occasions you were informed that the Executive Committee looked upon this matter with grave concern and that there was a possibility that the Executive Committee would take action to suspend your privileges and terminate your contract."

Averill had not been so informed, at any time, of pending action. The letter from the HMO completely avoids the discussion of ethics.

Day 999: The birth of the couple's first son. At this moment in time Dr. Averill is involved in teaching in a psychiatric residency training program in a country of the former British Empire. He is invited by the faculty there to coordinate the psychotherapy education for the residents in training, in appreciation of qualities that mark him as unusually conversant with the nuance and vicissitudes of therapy. He is also involved with a national effort to bring awareness to the epidemiologic proportions of mental disorders and the paucity of resources devoted to those in need. His efforts are recognized as contributing real value in the effort to improve the quality of mental health care.

Day 1027: Averill is informed by long term friends at the inpatient psychiatry unit, where he formerly held privileges, that a position is becoming available. Dr. Averill, in conversations with hospital administrators, is assured that they know of the circumstances of the events that led to the couple previously leaving the area. Senior psychiatric nurses and staff have come forward to validate that the couple's legitimacy has been greatly underestimated. Averill is assured that these events will not decide the hospital's willingness to employ.

Day 1037: A job interview is accomplished via an international phone call. Dr. Averill is informed that he is at the top of the list of candidates and is clearly considered the leading candidate by the headhunters' organization tasked to fill the position.

Day 1043: Dr. Averill receives a phone call from his counsel. The Medical Evaluation Board is planning to proceed to a statement of charges. It is the first contact in 683 days.

Day 1048: Christmas.

Day 1064: The headhunters' organization calls Dr. Averill to inform him that the administration at the hospital where he has interviewed had not been informed that there was an ongoing deliberation of the Medical Disciplinary Board, and under these circumstances they are unwilling to extend an offer of employment.

Day 1131: The Medical Disciplinary Board issues an interim order for a forensic psychological evaluation.

Day 1138: Dr. Averill completes an inventory of psychological tests for Dr. Yates, the forensic psychologist.

Day 1175: The first of four clinical interviews with Dr. Yates. Yates opens with a clear disclaimer that he is in the employ of the Department of Health and the Medical Evaluation Board. He then sets out with a tone that is demanding and derogatory. He states his examination could well result in the removal of Averill's license. He will be requiring a polygraph at some point. He has not concluded whether he will require a plethysmograph test of penile tumescence. He does not state that he is in possession of, and has read, all of the accumulated documents produced by Ms. Vaile at the Department of Health.

The psychologist begins a long and tedious process of questions, interrupting and controlling the flow of information to write down paraphrased versions of Averill's answers to questions. By the attitude conveyed, it is apparent that the psychologist is from the outset convinced that abuse has occurred and is single mindedly going after confirmation. He states that the laws of the state permit the mandatory use of polygraph testing for sexual offenders, which he states, he considers a correct characterization of what has occurred. (The exact wording of the Board's question to this psychologist was their concern that Dr. Averill might have "a mental disease or defect that would impair his ability to practice medicine." No mention of sexual offender status exists in the referral question.)

Time and again Dr. Yates asks questions that make large qualitative assumptions about the nature of the relationship between Averill and Clayton. He implies that it is a foregone conclusion that the only interpretation of events is to consider Averill's actions blatantly unprincipled. The grueling interrogatory style is suggested in the literature of forensic psychology as appropriate to the evaluation of criminals charged with sexual deviancy. At one point Dr. Yates uses this adopted stance to threaten Dr. Averill. Averill had stated that 'it is absolutely impossible for anyone to completely control the exact time, place and set of circumstances in which he or she might first be introduced to the person destined to become his or her spouse.' To this; Dr. Yates menaces "Do you really want me to write down that you're rationalizing? You know that you are rationalizing. Do you want me to write that down?"

Gradually Averill becomes convinced that Dr. Yates does not wish to entertain any set of ideas other than those he has predetermined as morally binding.

Days 1179-1183: Averill and counsel review the current American Psychological Association resolution regarding polygraph testing. Averill calls the American Psychological Association's Board of Social and Ethical responsibility and is referred to a woman in the office of the Science Directorate who is felt to have the best grasp of this issue. She is familiar with the review literature Dr. Averill quotes. She tells Averill that he has indeed grasped the fundamental concern of the APA in regard to polygraph testing. Averill is told "There is absolutely no justification for the further use of the instrument." He is informed that it does not meet the standards for educational or psychological testing. It is not a "valid" test. Further, the APA resolution (current) states unequivocally that the validity of the instrument is "unsatisfactory" and that the reliability of the instrument is "unacceptable" in that it routinely identifies innocent people as guilty, with a likelihood of one in six. Averill is also told that Dr. Katkin, a prominent researcher, testified before Congress (recently) stating that a psychologist is "ethically prohibited" from using methods that have problems of this magnitude with reliability and validity.

Days 1187, 1214: As in the previous interviews, Dr. Yates pursues questioning in a consciously repetitive manner. Averill offers to have the evaluation videotaped. Dr. Yates declines. Within an hour span Dr. Averill is asked four times how often he masturbates. It is clear that when Dr. Averill moves to speak of his particular understanding of ethical principles, the psychologist's pencil never records a word. At the beginning of session two, Averill speaks for 30 minutes about the ethical principles involved and how reasonable people could disagree about which of the principles, including a patient's right to make an autonomous decision, was most central. Dr. Averill offers the psychologist the right to review three dozen affidavits of friends, family and professionals willing to testify to the wholesomeness of the union. The right to speak with Ms. Clayton, after review with her, is offered. Dr. Yates replies that he doesn't have the time to either review affidavits or talk with Ms. Clayton.

Day 1237: Averill is seated in the empty waiting area of Dr. Yates office. There is no interfering noise and the conversation Dr. Yates is having on the phone is plainly heard. He is asking the other party, unidentified, about any statements from Dr. Katkin concerning the use of the polygraph. His tone is dismissive, as though the concerns of the APA are spurious, brought up in this context only to avoid the questioned testing. Once Averill is again in session with Dr. Yates, the psychologist reiterates the demand for the polygraph. Dr. Averill declines on the advice of counsel, appreciative that the current position of the American Psychological Association makes mandatory compliance with the test, ironically, absolutely unethical.

At this Dr. Yates is incredulous. He states that Dr. Doyle thought Averill was a liar. He ends with a quizzical statement to the effect that he cannot predict what the Board will do when someone refuses a polygraph examination.

Day 1302: Issuance of the psychological evaluation:

> "Based on his academic and professional history, his intel-
> ligence is, informally, judged to be above average. [Averill]
> responded to the MMPI-II in a manner similar to two
> types of clients. The first type includes persons who are
> experiencing significant psychological difficulties but are
> defensive and unwilling to admit them. The second group
> includes essentially normal clients who are able to manage
> the routine psychological stress in their lives ... His clini-
> cal profile on the MMPI-II was essentially within normal
> limits, indicating the absence of evidence of significant,
> acute or chronic psychopathology. His test responses sug-
> gest, consistent with the presentation during interviews,
> that [Averill] sees himself as morally virtuous and sensitive
> interpersonally and emotionally."

"Diagnosis:

Axis I:	V62.20 Occupational Problem, Complaint to the Disciplinary Board
Axis II:	V71.09 No Diagnosis
Axis III:	No diagnosis"

> "[Averill's] unrealistic appraisal and defensive stance re-
> garding his conduct in becoming romantically involved
> with [Clayton] indicates that there is a risk of his again
> developing feelings for a patient (or a supervisee) which he
> could not manage appropriately, resulting in a repetition
> of sexual intimacies and/or extraprofessional relationship
> that could be damaging to the patient."

Dr. Yates suggests:

- Prohibition from treating any female patient between 16 and 65.
- Prohibition from social contact with anyone Averill supervises.

- Supervision with a Board-appointed psychiatrist.
- Referral for treatment from a psychologist /psychiatrist experienced in treating health professionals who have engaged in sexual misconduct.

Dr. Yates' evaluation contains significant distortions, errors of fact and comprehension, as well as a remarkable series of conscious omissions about matters, including, but not limited to, 30 minutes of relevant discussion of patient autonomy. The slanting of material, with the expectation of gaining leverage in the processes of the investigation, is the motivation for a formal complaint submitted by Averill to the Board of Psychology concerning his behavior and report.

Day 1322: The State Department of Health issues a policy statement from the Medical Evaluation Board. The Board will hold Dr. Averill accountable, retrospectively to this document, as though it existed at the time of the events, Day 152.

> "Sexual Misconduct Statement and Policy of The Medical Disciplinary Board … 'I will come for the benefit of the sick, remaining free of all intentional injustice, of all mischief and in particular of sexual relations with both female and male persons …"

> "The report of the Council on Ethical and Judicial Affairs of the American Medical Association indicates that most researchers now agree that the effects of physician-patient sexual contact are almost always negative or damaging to the patient … To maintain the boundaries of the professional relationship, a physician should transfer the care of the patient to whom the physician is attracted to another physician and should seek help in understanding and resolving the feelings of sexual attraction without acting on them …"

Under the definition of patient:

> "The fact that a person is not actively receiving treatment or professional services from a physician is not determinative of this issue. A person is presumed to remain a patient until the patient-physician relationship is terminated ... Once a physician-patient relationship has been established the physician has the burden of showing that the relationship no longer exists. The mere passage of time since the patient's last visit to the physician is not solely determinative of the issues. Some of the factors considered by the Board in determining whether the patient-physician relationship has terminated ... include, but are not limited to, the following:

- Formal termination procedures.
- The transfer of a patient's care to another physician.
- The reasons for termination.
- The length of time that has passed since the patient's last visit to the physician.
- The extent to which the person has confided personal or private information to the physician.
- The nature of the patient's medical problem.
- The degree of emotional dependence the patient has on the physician. And,
- The extent of the physician's general knowledge about the patient.

> "Some physician-patient relationships may never terminate because of the nature and extent of the relationship. These relationships may always raise concerns of sexual misconduct whenever there is sexual contact."

The bibliography of the Statement and Policy references the Council on Ethical and Judicial Affairs, AMA article in JAMA, Vol. 266, no. 19, pages 2741-2745. The statement of the Board does not reflect that same article when it references the following qualification:

"While most researchers agree that sexual contact between patient and physician is potentially deleterious, it is important to note that most research has been based on patients who have initiated disciplinary action against physicians or on patients whom subsequent psychiatrists or therapists have identified as being harmed by the sexual contact with the physician. Patients not harmed by sexual contact with physicians may have escaped the attention of researchers."

"It is of course possible for a physician and a patient to be genuinely attracted or have genuine romantic attraction for each other; therefore, before initiating a dating, romantic, or sexual relationship with a patient, a physician's minimum duty would be to terminate his or her professional relationship with the patient."

Exactly when Ms. Clayton at last realizes that her competence, autonomy and integrity are inconsequential to the calculated position that the Board has adopted, when she realizes that there is nothing that she can offer that will be held in regard, she slowly and inexorably drifts into a severe Depression.

Day 1411: Averill ruptures a lumbar disc. An emergent cauda equine syndrome, with compression of nerves and nerve damage, prompts emergency surgery.

Day 1414: Christmas.

Day 1496: After long conversations with her husband, Averill, Ms. Clayton is psychiatrically hospitalized.

Day 1502: Present this date at a settlement conference are Averill and counsel, the principle lead representative of the Board (author of the policy statement), Ms. Zachary, the Assistant Attorney General for the State Department of Health, attorney for the Board and the physician consultant to the Board. Ms. Zachary opens the meeting and states that she has done a great deal of work in producing this Policy statement and that clearly something very wrong has occurred here. In the course of discussion, the availability of 18 immediately available opinions from psychologists, psychiatrists, psychiatric nurses, social workers, a lawyer and family member are referenced, to hopefully add balance to the deliberation. A discussion between the Attorney General and attorney for the Department of Health is overheard as the affidavits are derisively dismissed as "testimonials."

The documents, on the table, are not even looked at. Ms. Zachary turns to Dr. Averill and says that *no matter what he does, the Board is of a mood to find him guilty. That if he moves this to a hearing that is what will happen.* The Medical Ethics Consultant to the Board echoes that statement, in an aside during break. Aware that the only recourse to fair hearing will be reached at the level of appeal to the State Supreme Court, that at the same time Ms. Clayton is currently depressed and should not be subjected to additional stress, Dr. Averill and counsel move to agree to a Stipulated and Agreed Order.

Day 1536: The case is formally heard before the Medical Board. It is clear from the nature of the discussion that the members of the Board, with the exception of Ms. Zachary, are only remotely familiar with the case particulars. Ms. Zachary assures them that this is an appropriate level of intervention. The Order includes the following condition:

"The respondent shall provide written notice … that he is under disciplinary restrictions imposed by the State Medical Disciplinary Board …"

The agreed language of the notice, to be signed by all female patients subsequently, becomes:

> "I certify that I have been informed by [Dr. Averill] that
> he has entered into a Stipulation and Agreed Order with
> the State Medical Disciplinary Board … I understand
> that the Board has taken this action because [Dr. Averill]
> became involved with and is married to a former patient."

Day 1627: Ms. Clayton begins to recover from Depression.

Day 1921: On the occasion of the first scheduled review, the Board moves to terminate its jurisdiction in this matter, formally closing the issue. The reviewer for the Board shares that Ms. Zachary was alone in her willingness to pursue this issue; the other Board members did not find it deserving of the same disapprobation.

Day 4029: Dr. Averill writes to the Board:

To Whom It May Concern,

I am writing as a physician licensed in the state of [this state], concerned that you may be disadvantaged by an unfortunate design. I respect your integrity and the great service you provide for the people of [this state] in your role of monitoring physicians' practices. The issues that reach you are very complicated. Your steadiness and concentration to the task at hand help to determine that the quality of care delivered continues to improve. Thank you for that.

My letter to you is not written for personal gain. There is no issue before the Board for me at this time, nor is any anticipated. I have had a previous action by the Board [day 1536], jurisdiction terminated [day 1921]. This letter is not written to bring up those resolved issues for the sake of acrimony. I write out of concern generated by my participation in the University of [state's] course in Medical Ethics and the keen interest I have had in that subject area subsequently. My thanks to the Board, by the way, for recommending that course. It was in that course that my readings and review of the complexities of medical ethics broadened.

My concern is that the current methods of investigation and information gathering in matters of possible sexual misconduct do not reflect the type of review appropriate to ethical inquiry. The information getting to you is sufficient to allow adequate assurance of prevailing in prosecutions, but insufficient in that it may not allow you to [appraise] the ethical context of the matter at hand.

Before you dismiss this as too abstract, allow me to use my own experience to bring these thoughts into focus. My aim is not to unduly target criticism or underestimate the difficulty you routinely face in the issues brought before you. My experience is the only currency I possess that will admit me to this frank discussion. It is my hope that my experience might encourage you to review your information gathering tools. From Beauchamp and Childress, Principles of Biomedical Ethics;

"It seems mistaken, then, to say that ethical theory is not drawn from cases but only applied to cases. Rather, cases provide data for theory and are theories' testing ground as well. Cases lead us to modify and refine embryonic theoretical claims, especially by pointing to inadequacies in or limitations of theories."

The facts in my case are easily reviewed. I became involved with an ex-patient two weeks after terminating a short-term therapy. The issue was identified to the Board by a [very] jealous, separated husband, my employer and myself. The Board received this case in [date]. At that time, the only available statement of the American Psychiatric Association was to the effect that such a situation was *almost always unethical*, there was no automatic assumption that it was proven, a priori, by historical accounting of time lines or the presence of a doctor-patient relationship that preceded the involvement. The Board issued a Policy on this issue 33 months later; which is essentially unchanged in the (current) version. That Policy was, I believe, influenced by a change within the American Psychiatric Association's Ethics Committee, which, subsequent to (Day 152), drew a much more definite conclusion of the inevitability of abuse in these same circumstances*. I was held accountable to the Board Policy (drafted 33 months after the event) and was referred to a Forensic Psychologist for an evaluation of my danger to the public. The Forensic Psychologist found no mental illness but concluded that I should have restriction on the right to see female clients. The Board and I signed a Stipulated and Agreed Order for a period of probation, a fine and ethics education, based on a charge

---

* The Committee's opinion did not represent [a] consensus among psychiatrists. Notably, the opinion was not shared by several past presidents of the American Psychiatric Association, arguably aware that it constituted a retreat to the safety of a simplified notion of a complicated ethical dilemma.

of 'abuse of a patient'. At the one-year mark, when reviewed, the Board terminated its jurisdiction and closed the matter.

How was the advice to the MEB emanating from the investigators and prosecutors insufficient [and inconsistent] with regard to an ethicist's review?

1) The Board's representatives never spoke to the woman involved. Her statement to the Board was dismissed as irrelevant. She was clearly of the opinion that there had been no abuse.

2) The Board was not made aware of the opinion of the woman's psychologist, herself a member of the corresponding Board for Psychology in [this state], treating concurrently, who clearly stated the woman had not been abused.

3) The Board was not made aware that the woman's psychologist believed her to be competent. The investigators' presumption of incompetence allowed for the woman's perceptions to be dismissed as irrelevant.

Regarding competence:

> "Competence judgments thus have a distinctive normative role of qualifying or disqualifying the person, and the concept itself cannot be understood apart from this grading function. These normative judgments are sometimes highly contestable and yet are presented as empirical findings … Such a declaration, presented in the guise of an empirical determination of incompetence may conceal a value judgment about what a rational person would do that harbors an unduly narrow conception of rationality." Beauchamp and Childress, <u>Principles of Biomedical Ethics</u>.

4) The Board received a Forensic Psychologist's report with more than a dozen errors. The patterning of errors made the neutrality of the report entirely suspect. Among the omissions, the report failed to reflect the subject's

acknowledgment of [responsibility] in potentially placing an ex-patient in harm's way. That item, reflecting conscience and considered of pivotal concern in a Forensic enquiry, was dismissed as irrelevant. [Dr. Yates'] report was the subject of formal complaint to the [local] State Psychology Disciplinary Board, a matter on open record.

5) The Board was not placed in receipt of three dozen affidavits from four psychiatrists, ten psychiatric nurses, six social workers, family members, professionals and friends of the couple. These affidavits specifically commented on the nature of the relationship in question as not remotely reflecting abuse. The affidavits were quashed by state attorneys at conference, dismissed as irrelevant. Included in the dismissed material was an opinion from the woman's next psychiatrist to the effect that she was competent and had not been abused.

6). The Board was not presented material about the discourse of the involved parties to consciously deal with the ethical issues of their situation. That discourse was dismissed as irrelevant.

7). The Board was not advised re: the issue of the couple's religious/spiritual understanding of the underpinning of their relationship, also dismissed as irrelevant.

8). The Board received no advice from the Medical Consultant to the Board of a variance from accepted norms of ethical inquiry. Arguably the ombudsman of ethical review, his only seven words with the accused were to state "The Board intends to find you guilty."

What is wrong with this picture? The answer to that question depends on whether your aim is to write and effect policies that are airtight and proceed to conclusions of law in an unimpeded manner, or, alternatively, whether your aim is to have a policy that remains conscious of the field of medical ethics that you seek to ensure. The conflict surfaces in the selection of the phrase in the Policy relating to the relationship of physician and ex-patient. "The burden of proof will be on the physician to prove that no relationship existed." It is at that moment that the differing motivations and attentional qualities of ethicists and prosecutors are made most visible.

The prosecutors include this phrase to make the processes of the Board functionally equipped to capture the egregious behavior of marauding, exploitative, abusive physicians.

An ethicist, in contrast, would not assume that the existence of a relationship was proof in any material way of the presence of a marauding, exploitative or abusive quality. Specifically informed of mental health issues, an ethicist would know that the competence of every mental health patient is unique, every conceptualization and execution of therapy separate in its reliance on the development of dependence (some remotely, if at all), every situation standing alone in relation to attention devoted to the maintenance of patient autonomy. Fundamentally, an ethicist would not enter a situation with declarations of "fact" and then force parties to disprove them.

As it turns out, there is a very sharp disparity between the enforcement of law and the neutrality seen as an essential ingredient in the review of matters minimally necessary to comprehend the full development of an ethical analysis. If a case is routinely handed to a prosecutor, what will evolve, in accordance with prosecutorial ethics, is the trimming back of details to suit the adversarial advantage needed for the courtroom. An ambiguous picture will be made over to appear unequivocally, inescapably obvious.

Let me just weigh in here long enough to remind you that my own situation was not at all obvious. From the very same article quoted in the bibliography of the (Board) Policy:

> "...It is of course possible for a physician and a patient to
> be genuinely attracted to have genuine romantic affection
> for each other; therefore, before initiating a dating, ro-
> mantic, or sexual relationship with a patient, a physician's
> minimum duty would be to terminate his or her profes-
> sional relationship with the patient."

The drafters of the Policy omitted these caveats and presented their distillation as proof pertaining to their perception only. Their highly selective review of this and other available pertinent opinion disinclines the intention of a systematic review of ethical perspectives from which emerges an understanding of the total context, recommended by Albert Jonsen, Professor Emeritus of Ethics in Medicine. It is clear that they went to the literature to find support for their view, later evidenced as a notion that a law (or Policy) may be relied on to adequately describe the only possible interpretations of ethical principles in an infallible manner. Ethics recognizes this as a deontological theory. What the Board may not have been informed of is that reliance on deontological positions is espoused by a fringe element of medical ethicists, identifiable by the distinction between their radical position and the main body of prevailing opinion.

There is great trepidation among ethicists for the notion of relying solely on the law and its accompanying strong paternalism. The fear is that this strong paternalism can annihilate considerations of patient autonomy. From Beauchamp and Childress, Principles of Biomedical Ethics,

> "A policy or rule permitting strong paternalism in professional practice is not worth the risk of abuse [emphasis added] the policy or rule invites ... We have also maintained that paternalistic interventions are seldom justified because the right to act autonomously almost always outweighs obligations of beneficence toward the autonomous agent."

I take that to be a strong caution about rapidly dismissing a patient's right to determine [her] own course of action (including the ability of a woman, once a mental health care recipient, to understand what is at stake and competently terminate a doctor-patient relationship).

Within the field of medical ethics there is an awareness that reliance on law will most assuredly eclipse what passes for standards of ethical review. In my own case, an ethicist would not fail to appreciate the size and value of the facts that had to be deleted before a conclusion of law appeared obvious.

It is instructive to find that the woman involved, happily my wife of 9+ years, and I continue to live in a spiritually grounded relationship. We are devoted parents to our children, born on [day 999] and [day 2073]. We don't think very often about what happened to us in the machinery of the Board's process. The confident predictions that were raised about us have not materialized. We are an example of how a myopic conceptualized policy can intone narrowness and, in turn, misappropriate respect for the avoidance of an adequate review of ethical properties. That policy is completely unable to speak to the health of our relationship, completely at a loss to explain or describe it.

A theory concerning psychiatrists and ex-patients (that their involvement may be categorically eschewed), proven to have limited predictive value, can be defended indefinitely or amended. The appeal of that theory is the degree to which it appears to wrest certainty from ambiguity, absoluteness from the more human truth and its variations. I find it hard to divorce myself from the notion that theories must afford reconciliation with facts and not rely for their continuance on the annihilation of facts that are uncomplimentary.

I wish to make clear that there are significant problems with how the Board has been advised in the drafting of their Sexual Misconduct Statement and Policy. At a minimum, what is missing in this picture is a device that would alert the Board about those instances where automatic reactions on the part of investigators and attorneys are insufficient to understand the ethical [issues] identified or to ensure the rigorous review of competing ethical considerations.

I have a suggestion, offered respectfully. The Board would be well advised to hire or contract, in these matters identified for review, the impartial analysis of a professional ethicist. That would increase the likelihood that the Board would only become involved to the degree that their action is required.

I apologize for the fact that in order to conceptualize medical ethics issues it is necessary to use words in sufficient detail to draw comparisons between abstractions. That, I assure you, is less an issue of my personality than a real reflection of the literature. Thank you for your patience and steadiness in consideration of the merits of this letter.

Sincerely,
[Dr. Averill]

Day 4310: Dr. Averill concludes his comments to the Board:

To Whom It May Concern,

I previously have written to the Board concerning the absence of appropriate perspective in the approach taken by advisors to the Board concerning the development of evidence pertaining to the ethical dimensions of issues related to purported instances of abuse, specific to the relationship of psychiatrist and ex-patient. That letter is attached for your convenience. I trust you distributed that letter to members of the Board, I have heard no reply to date, although it is true that I did not ask for one. The only relevant change in that letter's content today would be to indicate that presently my wife and I are happy to announce our tenth wedding anniversary.

As part of the compliance with my Stipulated and Agreed Order in [day 1536], our agency handed a written notice to all the female mental health patients I saw, in capacity as psychiatrist, for the 13 months that the order was in effect. It impressed me greatly that, as I kept track, more than 80% of the women who commented on the notice told me in no uncertain terms that they were not surprised that the Board would hold such a prejudicial view concerning what they considered an ultimately objectionable sentiment; that the judgment of a former mental health patient can be summarily voided of all value; that an ex-patient can be removed without recourse from their autonomous purview to enter a spiritually grounded relationship, existing without any evidence of manipulation or deceit; that as an ex-patient, they would be frozen out, forever removed from the ranks of those people who had opinions that mattered simply because they had once graced a psychiatrist's door. They used this opportunity to instruct me that it was exactly this type of bigotry that they feared most in their lives; that people would blithely assume that, because they had once received mental health treatment, they were forever suspect and incompetent.

I assure you; I was surprised to find that these women instructed me, not in the wisdom of the Board's position, but in the reasons why strong paternalism has the capacity for great harm. That the theory so touted by the opinion referenced in my previous letter, championed at one point by the Ethics Committee statement of the American Psychiatric Association and upon which the Board took their instruction, had the inevitable effect to reify the stereotypes and prejudices, concerning mental health care recipients, that the APA otherwise publicly abjures.

These women taught me through their hard-earned insight that the position of the Board was not totally representative of the philosophical, rigorous honesty that is otherwise the signature of medical ethics, that any policy that results in a class of patients being systematically disallowed review by the normal processes of ethical deliberation is simply indefensible. Normal processes that look simultaneously at the areas of beneficence (paternalism), autonomy, patients' desires and justice. Normal processes that assiduously attempt to avoid jumping to conclusions. Normal processes that allow for balance and common sense to advise the weighting of the component aspects of review, specifically mindful that no single principle of Medical Ethics has such primacy that it vacates the others or automatically authenticates its own contribution as invariably [superior] to all other considerations. Normal processes that before the 1990's began to turn a corner toward an increased awareness of the importance of patient autonomy and to look with increasing cynicism at the historically misplaced overemphasis of paternalism in Medicine. Normal processes that are conscious of an expectation in Medical Ethics that each situation is unique, requiring a fresh look, and that entreaties to adopt concrete attitudes about the ultimate correctness of "strong paternalism" are dangerous departures from Ethics inquiries.

No one told [Ms. Clayton] that by entering therapy, she had been forced to enter "psychiatry" on the line immediately above "spirituality" when it came to important guiding principles of how to live her life. No one told [Ms. Clayton] that by entering therapy, she had to disclaim any possibility of trusting her own competent judgment concerning her ability to entirely understand what was at stake and consciously decide to terminate a doctor-patient relationship. No one told [Ms. Clayton] that by entering therapy, she had entered a world where the normative principles of Medical Ethics were absent, replaced with a popular ideology peculiar to the nascent development of ethics in psychiatry that bears exceptionally little resemblance to what passes for Medical Ethics, otherwise. Nor was [Ms. Clayton] warned that in these matters her sentience would be summarily dismissed.

In her mind, therapy was and is what a competent person employs to maintain the precision of judgment that allows for continuous, conscious autonomy. In her mind, she never surrendered her position of competence; she never ceded her autonomy; and the hurt, in reality as opposed to theory, that derived from this situation occurred exactly when the Board would not allow her to differentiate her competence, autonomy and happiness from their imagined offense. When the efforts of her psychologist and psychiatrist (receiving care on transfer) to document her competence, and the fact that she had not been abused, were consciously overlooked. When dozens of affidavits from professionals, family and friends were contemptuously dismissed at a settlement conference. At the exact moment when the Board drove home to her that she was inconsequential, that concerns for her were less important than the Board's need to prevail, she was truly and deeply victimized. Owing exactly to these circumstances, disallowed the opportunity to defend the honor and innocence of her relationship, she slowly succumbed to an unparalleled seven-month, severe Major Depression. Make no mistake. She resolutely sees you as the proximate cause of her injury.

If you were to talk to [Ms. Clayton], which you have not, you would find her a most intelligent human being. You would find her, even as her treating psychologist found her at the time of these events, a most logical and competent person. You would discover that there is nothing in her feminist nature to suggest that she would be satisfied with any agreement that would place her in a position of suffering her autonomy in the service of an ideology. You would find no indication that your narrowly defined definition of "abuse," floated upon theory alone, was in fact something that to this day she finds relevant or true for her in the least.

"Time will tell." "The truth will out." Sayings that remain in common parlance because they continue to transmit common sense. The truth that time continues to evolve allows for an appreciation of the inadequacy of your advisors' discernment.

So, at the end of the day, what you have, at most, is a pyrrhic victory. Yes, you have the power and therefore the ability to prevail in your assertion, with regard to [Ms. Clayton] and me, that your theory was, and remains, more important than our fact. You have demonstrated your ability to draft a Policy post hoc, declare it binding and arrange your fact-finding exercise to systematically disallow all alternative information. The cost of prevailing, in the exercise of your power in this fashion, is the loss of your ability to honestly claim that Medical Ethics supports either your methods or your conclusions. I am no great student of History, but I doubt seriously that you want that position.

In the present world, in fact, it is sensible to assume that the "burden of proof" passes to you to demonstrate that in this matter your theory, more enduringly than our spirituality, is the organizing principle that best affords total comprehension of these last five million uninterrupted minutes of a loving marriage. We hope it begins to make increasing sense to you that you were, and are, missing a vital component of what constitutes a Medical Ethics review. The expertise of an ethicist is not, through assumption, in the possession of reviewers in your employ, designing to furnish a legal result. I reiterate my call for the Board to hire the independent opinion of a professional ethicist to review matters of identified ethical concern.

[Ms. Clayton] and I extend you these earnest sentiments now, shortly following the occasion of our tenth anniversary. There are more things on Heaven and Earth than are dreamt in your Policy.

Sincerely, [Dr. Averill]

Day 8962: Dr. Averill, over the last 13 years, has contributed significantly to the community. He has held position as Medical Director of a community mental health center for 10 years, has the uniform respect of physicians and health care professionals in a disadvantaged county where he has worked, has been relicensed and has worked again for a brief time in the foreign country where his children were born, was the only psychiatrist hired by the [state's] Children's Administration to work in the history of [the state], was an Associate Professor, University of [state], School of Nursing, was hired by the largest HMO in the country at age 60, licensed by a second state, and has worked for two sovereign Indian Nation's mental health services.

Dr. Averill is, in the run up to day 8962, encouraged by colleagues who have worked with him, two physicians and a nurse, that the original HMO would very much consider him an excellent addition to the staff, and an offer to interview for a job is extended. It is 24 years from the original set of occurrences. Dr. Averill cautions that the previous history may yet have a lingering corporate animus, and the Psychiatry Department is advised to review the issue, before proceeding. Review within the department occurs, all issues on the table. The decision, from corporate levels within the department, is to proceed with interviews.

Interviews at the local level and corporate level occur and are universally positive. A recommendation for hiring Dr. Averill leaves the department for administrative review.

On day 8962, the HMO's Associate Medical Director for HR and Compliance does not endorse the rehiring. Workers at the local site are amazed and dismayed by the decision. Not so much, Dr. Averill.

Day 8984: Christmas.

CHAPTER TWO

ETHICAL ANALYSIS

Kangaroo Ethics

The trail that one travels in a medical ethics inquiry may turn away unexpectedly from an anticipated outcome. In the disciplined review of medical ethics issues, an overwhelming sentiment that precedes inquiry may be sufficiently counter weighted by the examination of simultaneously applicable principles, such that the conclusions reached are not what was anticipated. The discipline of an ethics inquiry will take slightly differing forms in different regions, dependent on custom and the education of the participant reviewers.

For our purposes, the rich case notes of this presentation allow a thorough review of the issue of abuse in the relationship of psychiatrist and ex-patient. We will start with the assumption that medical ethics issues are not simple. Unlike a jigsaw puzzle that will always be conferred in the same cast when successfully completed, we must move into the study of this issue without presiding prejudice. The picture before us needs to be reviewed with a series of lenses, developed and honed by the craftspeople of ethics. Only by slowly changing the focal length of the lenses at hand does the multidimensional picture form.

We will follow an inductive model of medical ethics decision making.[2] This accredited approach starts with the concrete particulars of the case developed in factual sequence. It elicits, through a review of dimensions of clinical decision making, all of the pertinent, applicable principles for the case at hand. An analysis of the areas of Medical Indications, Quality of Life, Patient

Preferences and Contextual Features will allow sufficient scope to play out the simultaneously applicable and competing purview of the principles as they weave the fabric of understanding from the threads of our discussion. This time-honored quality of medical ethics, which holds it impossible to get a complete comprehension without circumnavigating the identified issue with rigorous discipline, is an essential ingredient of our current effort.

My hope is to create in this analysis an unbiased atmosphere where the properties of the operative principles may be considered in a way that permits each to be entertained as though they are primae facie binding.[3] Defended by Beauchamp and Childress throughout their *Principles of Biomedical Ethics*, the issue of primae facie binding holds that moral rules and principles are binding, but not absolutely binding.

> "A composite theory permits each basic principle to have weight without assigning a priority weighting, or ranking which principle overrides in the case of conflict, with dependence on the particular context, which always has unique features."[4]

These authors argue that by invoking this capacity (primae facie binding) in the analysis of principles and rules, there is allowed a dimension in which to consider that a single principle may suggest two competing and relatively equal alternatives. It is a strong encouragement to remain balanced while examining evocative topics.

Our efforts will forsake the alternative deductive model of medical ethics analysis. A model of deductive design holds that you start an analysis with a "correct" theory that elicits the apropos principle involved. The "correct" theory then suggests a rule for a set of circumstances, and the rule is unassailable.

Increasingly in the definition of Medical Ethics throughout the twentieth and into the twenty first century, there is a cumulative perception that the perils of deductive models of review recommend our alternative, inductive approach. The intention of inductive processes is, at its best, to act as an antidote for bias, as the progression of a formal ethical analysis is developed.

Medical Indications

The first of our points of review will be the area comprised by Medical Indications. In this portion we will review the diagnosis of the patient, with particular care to distinguish the competing formulations that were held by the primary characters. The prognosis of this medical condition, as conceptualized, will then be reviewed. The efficacy of treatment, the utility of treatment and the goals of therapy are then reviewed with particular attention to the principles of medical ethics that are most applicable, beneficence and non-maleficence.

Diagnosis

The American Psychiatric Association has evolved a considerably sophisticated official nomenclature in appreciation of the diversity of clinical presentations and the multiple dimensions in which they may be simultaneously described. Their efforts created the multiaxial diagnoses with the 1980 publication of the *Diagnostic and Statistical Manual of Mental Disorders,* 3rd Edition. This multiaxial system, in widespread clinical use at the time of these circumstances, as DSM-IV, where it was described:

> "A multiaxial system involves an assessment on several axes, each of which refers to a different domain of information that may help the clinician plan treatment and predict an outcome. There are five axes included in the DSM-IV multiaxial classification.
>
> | | |
> |---|---|
> | Axis I: | Clinical Disorders |
> | | Other Conditions That May Be a Focus of Clinical Attention |
> | Axis II | Personality Disorders |
> | | Mental Retardation |
> | Axis III | General Medical Conditions |
> | Axis IV | Psychosocial and Environmental Problems |
> | Axis V | Global Assessment of Functioning |

"The use of the multiaxial system facilitates comprehensive and systematic evaluation with attention to the various mental disorders and general medical conditions, psychosocial and environmental' problems, and level of functioning that might be overlooked if the focus were on assessing a single presenting problem. A multiaxial system provides a convenient format for organizing and communicating clinical information, for capturing the complexity of clinical situations, and for describing the heterogeneity of individuals presenting with the same diagnosis." (P. 25, DSM IV)

"...there is no assumption that all individuals described as having the same mental disorder are alike in all important ways." (P. xxii, DSM IV)

Within the wealth of case notes presented in the initial section of this book, it is important to remember that certainly not every element of life that involved Ms. Clayton at the time of the events chronicled was in specific reference to an Axis I Clinical Disorder (i.e., Bipolar Disorder.) In fact, by the then current standards of DSM III-R (in use by that time), Ms. Clayton's multiaxial diagnosis would have been recorded:

Axis I	Bipolar II Disorder, Not Otherwise Specified, in remission
	Dyssomnia, Not Otherwise. Specified
	Uncomplicated Bereavement
Axis II	No Diagnosis on Axis II
Axis III	Hypothyroidism, in control on medication
Axis IV	Psychosocial Stressors; Problems with marriage, loss of mother
	(predominately enduring circumstances) Severity, severe
Axis V	Current GAF (Global Assessment of Function): 75
Highest	GAF past year: 75-80

The sophistication of this approach to diagnosis allows the individual circumstances to be placed in perspective. For example, does it follow that in this complete diagnostic picture, a sleep difficulty reported by Ms. Clayton is solely attributable to a rampant Bipolar Disorder? No, it does not. Is it clearer why Ms. Clayton had an immediate improvement in her sleep when she moved into the house of friends, away from the marriage? Yes, that might be an expected consequence of living in a less stressful circumstance. Seeing the whole story at a glance, a sleep disturbance is understandable as a reaction to the Axis IV stressors, improving as they are mitigated.

One entire set of difficulties emanates from the lack of comprehension of the necessity for this devotion to detail. To relate to the history of Ms. Clayton as though every element of her life and every symptom she experienced was a reflection of Bipolar Disorder is exactly what DSM III, III-R and IV (indeed, even DSM 5) caution against. The APA considers oversimplification misleading, that without a full psychosocial appreciation of the person and his/her life circumstances, the picture is predictably distorted by trying to condense disparate elements into a one-line diagnosis. People who conceptualize mental disorders as unidimensional are more prone to mistake the causation of symptoms. More prone, because they have underestimated the simultaneous effects of complimentary Axis I, II, III and IV diagnoses. There is a tendency to over read the symptoms that are present as evidence pertaining to the identified clinical condition, not recognizing that they may emanate from an entirely different source.

In our case noes, it becomes clear that Mr. Eason paid little heed to the contributions of bereavement and marital difficulty as reasons for a sleep problem. Although in any other person, it might be normally expected that sleep would be difficult following the death of one's mother and disintegration of one's marriage, instead Mr. Eason used this issue to inflame sentiment against Dr. Averill, as though Dr. Averill was obviously taking advantage of a woman in the midst of a severe manic episode. Mr. Eason's evidence hinged on the etiology of that sleep disorder as invariably owing to a mania.

The medical record for Ms. Clayton portrays the most salient appreciation of her history prior to the events in our narrative as consistent with the Diagnostic Statistical Manual definition of Bipolar II Disorder, Not Otherwise Specified, In Remission. In this recognized mood disorder, there are alternations of periods characterized by slightly increased energy, with contrasting periods of more severe dimensions of Major Depression. Often the time between episodes is interspersed with prolonged periods of completely normal moods, uniquely developing on an individual basis, with great variability.

A distinction is drawn between this condition and Bipolar Affective Disorder, type I. In the latter, type I, there are periods of mania (as opposed to periods of hypomania.) The qualifying difference in the DSM III indicates that to qualify as a mania requires

> "...a mood disturbance sufficiently severe to cause marked impairment in occupational functioning or in usual social activities or relationships with others or to necessitate hospitalization to prevent harm to self or others."

Our review of facts from the case point to evidence from senior psychiatrically trained nurses on a psychiatric inpatient unit, who validated directly the continuous competence of Ms. Clayton, at all times during this critical period leading into the time line of the case narrative. Moreover, the statement of the psychologist and subsequently treating psychiatrist who received the case on transfer from Dr. Averill reflects an awareness that the criteria for mania are not met. Two close friends, into whose house Ms. Clayton repaired at the time of her separation from Mr. Eason, similarly depict her as displaying considerable ability to observe, concentrate, abstract, reason and effectively communicate ideas throughout this period, in a manner inconsistent with a definition of impairment.

The distinction is important. People without clinical training may make qualitative judgments about the presumptive conclusion of incompetence based on stereotypes or prejudices particularly owing to their misunderstanding of Bipolar Disorder. In our example, it is likely this consideration needs to be remembered when considering the actions of the Board and their surrogate, Ms. Zachary, for purposes of creating policy for the general medical community. None of the Policy drafters had particular training or expertise in psychiatry. A person not trained to the nuance of this matter may erroneously assume, for example, that a Bipolar Disorder invariably confers a situation at the far extremes of sanity, quite possibly a psychosis, mistakenly assuming the diagnosis undoubtedly renders that person unable to meaningfully interpret the nature of events in which they participate.

The truth is far more complicated than that. People with a diagnosis of Bipolar II are frequently among those inventors, composers, authors and scientists who have given the greatest gifts to civilization. The presence of a label that depicts an evanescent mood disorder, sometimes present and often absent, does not disqualify the insight or virtues of the person so described. Nor is that label so invariably predictive of mental status at a given moment in time, such that direct evidence of mental status is of less importance in determinative discussions relating to competence.

In our case, the cumulative picture that takes form of Ms. Clayton is supported by serial professional opinions that are remarkably similar, all relating to her continuous clarity and competence. Her sleep disorder is qualitatively not to the threshold of recommending a more severe diagnosis. During the evolution of this case, in the heat of anger and despair of separation, Mr. Eason and Mr. Ott represent this issue as one of incontestable proportions, that Ms. Clayton is in the throes of a severe mental illness and unable to exercise judgment. Although there may have been many laudable reasons for their protective presumptions, there is no evidence of a professional nature that their depictions were timely or correct. Nevertheless, they had sufficient influence in this matter to elicit a cascade of harmful and mistaken assumptions. Mr. Eason held no formal mental health credentials.

One of the curious ironies of this case is closely related. There are some periods of mild elevation of a mood during which people with or without psychiatric diagnoses experience a remarkable clarity of thought and insight. In such a buoyant period, a person may be very content, creative, witty and charming. Far short of any dimensions of disturbance that would recommend a psychiatric disturbance, these periods often confer a degree of perspective, previously without parallel. It is entirely possible that in such a period of clarity, as opposed to mental illness, Ms. Clayton was able to see her husband's need to pathologize her behavior for the first time in stark relief. It is possible that the energy infusion associated with this moment, a non-clinical event, was interpreted, per Mr. Eason's developed custom, as evidence of mental illness. Harder perhaps for Mr. Eason to accept than the possibility that his wife's realization was accurate.

Prognosis

The prognosis of mood disorders is highly variable. The milder variants of mood disorder may have presentations that differ considerably from the statistics associated with more typical presentations of major mood disorders (i.e., bipolar disorder and major depressive disorder). Each individual with mood difficulties will describe his/her own unique time line of mood events. Although in general it is true that Bipolar Disorder tends to become more brittle with the passage of time, with more episodes of mood disturbance and less time between mood events, this statistical truth is in no way predictive on an individual basis of the likelihood of health versus disease states.

Efficacy

The efficacy of mood stabilizing medicine for Bipolar Disorders, recurrent Major Depressions or to potentiate antidepressant strategies is well established and beyond the scope of this review. Importantly, in our case, the psychotropic medication Lithium was prescribed for its ability to prevent a decomposition from a relatively mild sleep disturbance to a more damaging hypomanic episode. Lithium takes four to five days to equilibrate in a stable serum concentration. Indeed, one of lithium's chief difficulties as a psychotropic medication is the delay that clinicians and patients experience from the first dose to the emergence of clinical effect. Often used to help people in manic states, this medicine will require one or two weeks to begin to ratchet down the mood elevation. This quality has eventually led to the adoption of ancillary medications to adjunctively bring the mania to heel on a quicker time line.

So, it is true that a person who recovers from a sleep disturbance on the first day of lithium's use is, with a probability approaching certainty, not recovering due to the ameliorative effects of lithium. When Ms. Clayton moved to the home of friends, she was immediately relieved of her sleep difficulties. This piece of her recovery was not owing to the efficacy of lithium. It was owing to the diminution of her stress, directly.

Questions Dr. Averill had about the need for lithium were understandable. Lithium has the potential for side effect profiles in kidney and thyroid, skin, eliciting tremors, creating thirst and increasing urination. It has a narrow therapeutic index and can accumulate to toxic levels relatively easily. It was recommended at a time when the main proponent of the theme of extant mental illness was Mr. Eason, by Mr. Eason. It was not something that Ms. Clayton originally wanted at all. That Dr. Averill sought to refer Ms. Clayton to another provider for a more objective read of the need for lithium is laudable, given that he had reason to believe the illness aspect had been overestimated considerably.

The efficacy of therapy for the resolution of grief is widely accredited in the field of psychotherapy. The particular model of therapy employed was Thomas Mann's Time Limited Psychotherapy technique; where as much as possible the patient's autonomy is encouraged throughout, from the moment of establishing the patient's role in defining the primary focus. Completing the work on a time schedule is inherent in the design. Curiously it is not a therapy that fosters dependence, nor elicits unconscious or preconscious mechanisms. It does not require an analytic stance, the development or working through of transference. As a therapy technique, it is practiced best with a light touch, allowing the patient to gradually assume the momentum and lead in processing the time limit and its implications. When the therapy proceeds well with a highly motivated, mature personality, autonomy of the patient becomes a recognized component of the final pieces of therapeutic work, leading to a successful termination. A more mature person takes their leave from therapy, consciously.

Again, the distinction is important. Therapy is not a unidimensional event where all therapies create the same dependent state in the patient. But that is often the perception of the public, that a person in therapy with a psychiatrist has undoubtedly been placed in a position of extreme vulnerability, that psychiatrists may choose to manipulate patients in their dependence to the service of the psychiatrists' self-gratification. That there exists an uneven and manipulative aspect to a psychiatrist-patient relationship, invariably. This same proclivity would extend equally to all therapies in a blanket manner. The rules that require strict prohibition of relationships between therapists and ex-patients are an attempt to deal with this ubiquitous quality of the doctor-patient relationship, that psychiatrists are somehow inherently more dangerous because they have access to information and have knowledge of unconscious and preconscious defense mechanisms that gives them an unfair advantage.

This argument is a lynchpin in the development of the idea that there is an unassailable accuracy to the view that there invariably exists manipulation, deceit and self-serving in the relationships that follow an original design of psychiatrist and patient. This argument does not consciously allow for the distinctions between therapies, which vary in important capacities with regard to dependence and the capacitance for manipulation. This argument paints all therapies with the same brush, as though the details of the specific therapeutic approaches are unimportant.

The Principle of Beneficence

The component of medical ethics review that is the principle of beneficence is anchored to the ethic of Hippocratic medicine from the fifth century B.C. The most cited reference is "Be of benefit and do no harm." Hippocratic tradition is recognized as one of prudent paternalism.[5] In its most general form, the principle of beneficence asserts an obligation to help others further their important and legitimate interests. The obligation to confer benefits and actively prevent and remove harm has specific import in a biomedical context. Equally consequential is the obligation to weigh and balance the possible good versus harm of any action.[6]

Beneficence, as it relates to our case presentation, was perceived in two wholly separate and simultaneously competing contexts. As principles are not the property of any single perspective, it remains for us to delineate these contending applications.

From the perspective adopted and at the heart of the sentiments of the HMO, the investigator from the Department of Health, the forensic psychologist and the Medical Disciplinary Board, there would appear to be no disputing that the fundamentally correct approach to this situation is to assume that the principle of beneficence here would refer only to the fact that a psychiatrist would never become involved with an ex-patient. Here, the fact that a therapy occurs with a vulnerable human being who arrives for that purpose because they are emotionally unbalanced would strongly imply that the only appropriate caution to prevent harm would be denying any attraction that might arise. From this perspective any attraction is assumed to be powered by unconscious mechanisms of countertransference. These unresolved feelings are displacements from other unresolved areas of the therapist's personality. They are representative of defenses that require work on the part of the therapist to resolve, not act out in the relationship with the patient. Any therapist who does not agree with this assignment of the underlying mechanisms and importance of countertransference is inviting depiction as a poorly trained, and therefore, dangerous clinician.

From the perspective of the patient (beneficence and the obligation to do no harm, the obligation to further the important and legitimate interests of the patient) the meaning of this principle and how it applies is something entirely differently. Here we find the interests of a mature woman who values her independence and ability to know her own heart. Here we find a person who came to therapy not to be dependent on anyone, but to have an appropriate forum to discuss the next organizing of her adult understanding of a grievous loss. To this patient, the function of beneficence informs her therapist to never stand between her autonomy, her free choice and the most intact, spiritually informed person she is capable of being. To her, beneficence is manifest in the respect from the therapist to understand that her autonomy had never been taken from her. That, to do no harm, one would have to be conscious of her capacity to determine that she was healthy and informed at the moment she chose to terminate the doctor-patient relationship. Further, that this principle of beneficence does not solely relate to the doctor's role in isolation, but must allow the truth that her life was not fully comprehended, nor could it ever be wholly conceived in its conscious entirety, by the language of psychoanalytic tradition, encapsulated and reduced to a collection of defense mechanisms.

The actions of the psychiatrist were complicated by this dual reality. Trained in several forms of therapy, psychodynamically oriented, time limited psychotherapy, family therapy and solution-oriented therapy, the therapist had experience with forms of therapy that do, and then again do not hold out the importance of these analytic interpretations. Not all therapies do, in fact, hold the significance of transference and countertransference to be pivotal, or even necessary, to the management of a successful therapy. In our case example, the psychiatrist conducted a therapy that allowed for the continuous presence of those personality skills already the possession of the patient, reminding her of those qualities as she solved her own dilemma of moving through the issues of grief.

Some of the power of the argument for first version of beneficence, as though it is only understandable from the perspective of the Medical Disciplinary Board, derives from the ability of proponents of this view to describe all therapies through the interpretation of the analytic versions of therapy, taking for granted that this is a reasonable decision. Quietly and confidently, they assume that other therapies are insufficient, in and of themselves, to understand how humans operate. More important, without publicly displaying the explanation of their assumptions or their implications, they move to depict normal human behavior in the language of defense structures and illness, giving no voice to possibilities of spiritual growth or the place of transcendent values informing life decisions.

In fact, less than 1 percent of all practicing therapists are analytically trained. The average length of time spent for a client in psychotherapy in the United States in the decade of our case presentation was between four and seven sessions. It is an anachronism to assume that therapy today relies on the development and working through of transference. It is a testament to the historic organizing power of analytic theory that so many people yet make the blanket assumption that the language and conceptual framework of these analytic theorists has been raised to the status of fact.

A curious buried artifact of *paternalism* is visible here. A sense conveyed that no one understands the patient better than the doctor. That the doctor's sense of the balance of defense mechanisms may be artfully substituted for the patient's sense of spiritual and individual correctness. The early psychoanalytic contingent trained in an epoch of medicine where paternalism held sway to a much greater extent than currently evidenced. Doctors in Vienna in the late 1800's were encouraged not to even tell patients of the details of their disorder or disease. They were encouraged to do everything in their power to promote a sense of cheerfulness and optimism, even at the expense of honesty, even at the price of deception. They took their tone from the Hippocratic tradition:

> "A doctor shall give necessary orders with cheerfulness and serenity, turning his attention away from what is being done to him; sometimes reprove sharply and emphatically, sometimes comfort with solicitude and attention, revealing nothing of the patient's future or present condition."

So, it was an integral part of the culture at the time of the origination of analytic theory that the physician was qualitatively separate, above the patient and bound by honor to remember that disparity in all clinical matters. This allowed the silent partner to this noblesse oblige to escape notice, the automatic rejoinder that to be in a relationship with a physician for the purpose of therapy, one had given over all future claim to competence, particularly if, as that patient, you saw something different in your actions than what the therapist was willing, or able, to conceptualize. In this moment, when the position of beneficence is tied so tightly to a version of strong paternalism, the speed with which a patient's capacity for competence is removed from sight is not unlike the move of a magician to draw attention to his right hand while his left palms the hidden card to a pocket. It happens so fast that you have trouble seeing it, even if you are looking for it.

The threads that comprise paternalism do not exist such that they can ever remove themselves from the fabric of autonomy. They are bound. A movement in one aspect results in a change in the equilibrium of both. Although it is not as commonly discussed, respect for autonomy is an alternative manifestation of the principle of beneficence. It requires an adoption of humility to conceive of it. It requires a relinquishing of the patterning of strong paternalism, so familiar to psychiatry. But it is a compelling, competing interpretation of beneficence that does not occur at the cost of patient autonomy.

As close as the facts of this case presentation will allow, Ms. Clayton and Averill understood these ethical issues and had lengthy discussions about them. Their actions were labeled as unethical by people enamored of one particular view of this issue of beneficence. But as we continue to follow the outline for a complete ethical review, we will visit the limitations of these criticisms.

Paternalism, and its ascendency to a position of *strong paternalism*, where it may trump all other simultaneously applicable principles, can be justified only if:

1. the harm prevented from occurring, or the benefits provided to the person, outweigh the loss of independence and the sense of invasion caused by the interference,
2. the person's condition seriously limits his or her ability to choose autonomously, and
3. the interference is universally justified under relevantly similar circumstances.[7]

These paternalistic actions are seen by ethicists as appropriate in health care only if:

a. a patient is at risk of injury or illness,
b. the risks of the paternalistic action to the patient are not substantial,
c. the actions' projected benefits to the patient outweigh the risks,
d. there is no feasible and acceptable alternative to the paternalistic action,

    e.   the infringement of the principle of respect for autonomy is minimal, and

    f.   the action involves the least infringement necessary in the circumstances.[8]

Infringement on respect for autonomy is considered in situations to be warranted only when "vital autonomy" interests are not at stake. For example, if a Jehovah's Witness refuses a blood transfusion because of a deeply held conviction, a vital autonomy interest involving spirituality is held to be at stake, and the infusion may not be given against the will of the patient.

In our review of this case, we have no evidence that the marriage of Ms. Clayton to Averill produced harm. There is every evidence that her capacity to choose autonomously was consistently intact at all times in the evolution of her relationship with Averill. There is no universal justification for blocking a marriage, held by the participants to be the highest expression of their mature spiritual life. The risk of paternalistic action is painfully evidenced in the etiology of Ms. Clayton's Depression, which followed her sense of not being allowed to differentiate her true experience, not being afforded the luxury to be respected in her competence, autonomy and ability to establish herself as evidently in capacity to understand the circumstances and to perform an informed choice. There were no benefits to Ms. Clayton of the Board's decision to move to prosecution. The infringement of the Board on her autonomy was annihilative. Their action was not the least infringement possible in these circumstances. Overall, the action to move the issue of this relationship through the processes of the legal arm of both the Department of Health and Board, without once talking to Ms. Clayton, does not satisfy the criteria suggested as the threshold for the intervention of strong paternalism by standards recognized within medical ethics. The normative processes of ethics were seriously abridged, with disastrous results. The vital autonomy interests that may have been present (in the couple's strong Buddhist belief that they had a fundamentally spiritual connection) were thoroughly ignored. Although reported to the forensic psychologist, spirituality was never mentioned in his subsequent report. It would be unconscionable for a medical ethics review to refuse to consider a religious or spiritual view that inclined to action, as though it was insignificant and not material to the deliberation.

From the President's Commission for the Study of Ethical Problems in Medicine:

> "The primary goal of health care in general is to maximize each patient's wellbeing: however, merely acting in the patient's best interests without recognizing the individual as the pivotal decision maker would fail to respect each person's interest in self-determination … A competent patient's self-determination is-and usually should be-given greater weight than other people's views on that individual's wellbeing. Respect for the self determination of competent patients is of special importance … The patient (should have) the final authority to decide."[9]

Quality of Life

> "The most fundamental goal of medical care is the improvement of the quality of life of those who need and seek care."[10]

Quality of Life is one of the defining dimensions within an inductive process of ethical inquiry, bearing directly on the subjective experience of the person involved. It is most often conceptualized in the context of deliberating over the utility, or futility, of continuing care for people who are terminally ill or actually drawing near to the moment of death. The area of sensibility that is illuminated by considering quality of life is not commonly envisioned in ethical considerations of psychiatric patients. At any one time, the World Health Organization estimates that somewhere around 16-17 percent of people qualify for a mental health diagnosis. The sad truth is that stereotypes still prevail in the public awareness concerning elements of mental illness - stereotypes exist that make the ready dismissal of considerations regarding quality of life issues all too palatable.

In our case, it is peculiar that so many people came forward to portray what they took to be an accomplished fact, that Ms. Clayton could not be trusted to have a clear idea of what animated her heart and soul. In this collective act of ignoring, several biases are visible.

There is the bias that somehow, due to the nature of having been a patient, Ms. Clayton was incapable of being healthy. That the initiation of therapy marked a point in time where the possibility of her awareness ceased. That in crossing that line into the therapist's office, she was, or should have been, aware that the therapist was in some tangible way capable of eliminating, without explanation, her capacity for judgment regarding her ability to discern a spiritual truth.

Mental illnesses are not created equal. Nor are they the same event for the same individual at different times in their life. If the presence of a mental health diagnosis, which may be in remission, is sufficient to automatically remove all hope of competence in understanding the position of spiritual matters in our lives, then a very large number of us are doomed to be directionless.

Among the issues that describe quality of life, a person's ability to address the question of spirituality is very important. The core of a person's spiritual beliefs is not determined by the color of skin, age, sex, job or past psychiatric diagnosis. The fundamental transcendent nature of spiritual experience is shared in all corners of the world by many differing faiths, all attempting to describe spirituality through different languages. Some are concretely drawn; some are more abstract. But the right to pursue the question of the meaning of life is equal in us all.

People with mental health diagnoses, which may be in remission, have this same interest of vital autonomy in pursuing a spiritual connectedness. They do not shed this human drive for quality within their lives because they once had a diagnosis. It is true that people in the midst of severe psychotic reactions are often delusional in the direction of material that presents with a language of religiosity, as they struggle in their terror to hold on to any deep-rooted explanation for the implausible experience of reality to which they are subjected. Psychotic disorders are but a tiny end of the continuum of psychiatric presentations, far afield of the often highly functioning people who have had the momentary displeasure to become anxious, depressed or to have crossed the threshold for a mental health diagnosis.

In our case study it required only an allegation of the presence of a debilitating mental illness to set in motion a series of events that fell like dominos. Once Mr. Eason presented Ms. Clayton as suffering a mental illness, the advisability of a strong paternalism appealed to the people he contacted. The issue of danger to Ms. Clayton gave rise to the immediacy of a protective reaction in those people first introduced to the issue of her relationship to Averill through the ministrations of Mr. Eason's attempts to inform, certainly with the aim of eradicating what he perceived as a medical menace. The power of the allegation and its recruitment of affect was sufficient to allow people to directly dismiss the notion that Ms. Clayton was in capacity to understand fully what was and what was not transpiring in this moment. The bias against those with mental health diagnoses is so strong that merely an allegation may stand for and become irrefutable proof, fueled by emotions operating at a fever pitch.

The consensual professional opinion of Ms. Clayton at the time was different. Suffering from a grief for the loss of her mother, unable to sleep because her husband was treating her without a modicum of respect, she was not in the throes of a severe mental illness. She was in the throes of a severely testing period, and according to professional psychiatric and psychological opinion, doing a reasonable job of acquitting herself.

There is no compelling evidence in our case study to discount the relevance of considerations regarding quality of life. A person in a failing marriage with trouble sleeping is not so removed from reality that she is incapable of discerning the presence and importance of a spiritual event. Ms. Clayton retained the right to her appreciation of spirituality informing her decision to be involved with Averill. Those interested in considerations of Quality of Life would not lightly suffer the implication that a diagnosis in remission necessarily disinclines or replaces spiritual judgment.

Patient Preferences

Competence

"In a medical setting the patient's capacity to consent or to refuse care requires at least an ability to understand relevant information, to appreciate one's medical situation and its possible consequences, to communicate a choice, and to engage in rational deliberation about one's values in relation to the physician's recommendation about treatment options."[11]

"There should be a moral presumption in health care parallel to the accepted legal presumption of an adult's competence to make decisions. The burden of proof would then fall on those who believe that a particular adult is incompetent to decide, and it would be necessary for health care professionals, often by appealing to the courts, to establish a patient's incompetence in order to disqualify him or her from decision making."[12]

"Competence judgments thus have a distinctive normative role of qualifying or disqualifying the person, and the concept itself cannot be understood apart from this grading function. These normative judgments are sometimes highly contestable and yet are presented as empirical findings. For example, a person who appears irrational or unreasonable to others might be declared incompetent so that treatment can be provided against his or her wishes. Such a declaration, presented in the guise of an empirical determination of incompetence, may conceal a value judgment about what a rational person would do that harbors an unduly narrow conception of reality."[13]

This area of ethical review commonly examines the applicable dimensions of informed consent, the capacity to choose, the refusal of therapy, the involvement of surrogate decision makers and the pivotal position of considerations of autonomy. The case notes allow for an examination of these dimensions in some detail.

There is every evidence that Ms. Clayton entered a long and detailed discussion with Averill about the possible import of transference and countertransference. Traditional ways of viewing the situation were most certainly discussed and examined. The 'appearance' of an act of harm was thoughtfully distinguished by Ms. Clayton from an 'actual' act of harm. The guiding principle affecting this insight in Ms. Clayton was the complete and integral spiritual understanding that, in odd circumstances, she had been introduced to exactly that person that she understood as her mate in this life. This allowed her the space to rationally consider the possible harms, to weigh the likelihood that her actions (as well as Averill's) were disingenuous, and to decide that her instinct and intuition informed her correctly.

Did Ms. Clayton have the right to terminate therapy? Here the power of the analytic presumptions come immediately to the fore. In every other field of medicine, it is held that a competent individual has the right to refuse therapy. A competent adult may decide that his or her interests are best served by ending treatment and terminating the doctor-patient relationship. There are no formal procedures now, and certainly none existed at the time of the events in the case notes, for informing patients of psychiatrists that their judgment to terminate a doctor-patient relationship would be regarded as less salutary. It is reasonable to assume that Ms. Clayton never imagined that by entering therapy she had forsaken her ability to leave therapy under her own power.

Are there outstanding reasons to believe that the existence of a psychiatrist-patient relationship is primae facie evidence against the right of the patient to terminate a relationship? Here the arguments of the analytic theory propose that unconscious mechanisms power the processes of therapy with such vitality that they may not be retired. Similarly, the emergence of affect in the process of a therapy is itself a springboard for further exploration into the processes of transference and countertransference. The appropriate stance is to never entertain the possibility of sincerity of these potential emotions, but rather to recognize them and work through them in supervision with an analyst or in therapy with an analytically trained, seasoned psychotherapist. As the argument against the possibility of sincerity of attraction is built, all remaining images of such relationships are portrayed as damaged psychotherapies, created by damaged psychotherapists, creating damaged patients. No exceptions.

What we have in this case is something different, requiring sobering reflection.

Here we have a healthy marriage of fourteen years duration, above the average for a marriage in the United States at present. We have no damaged patient.

Here we have a psychiatrist who in 41 years of licensure has had no boundary problems identified for review in his professional practice, with the exception of this case. We have a psychiatrist who has supervised therapists, who has coordinated a psychotherapy training program for residents in training. We have no damaged therapist.

We have a situation where to this day Ms. Clayton believes the therapy was conducted and concluded correctly, and then her life moved on.

Is it unduly narrow to consider that a woman entering therapy is immediately irrational if she considers that something more than the descriptions available through analytic theory is afoot? What if that woman does not endorse psychoanalytic interpretations and theory?

If the evidence of outcome is any measure of the binding validity of a theory, that holds that a psychiatrist-ex-patient relationship is fundamentally flawed, then we have in our case notes evidence that analytic theory is insufficient to describe the facts as they occurred. The presumption of Ms. Clayton's incompetence in this theory, which functions as its battery, is an attempt to discredit the judgment of psychiatric patients without review.

Contextual Features

The last of our major areas to be considered in an inductive model of medical ethics review is the contribution allowed by a review of relevant contextual features. This is where the relevance of benefits, harm and the rights of others are examined as they pertain to and are affected by clinical decisions. The interests of other parties and the general public interest are reflected upon in the medical decision making. Confidentiality and the protection of the public are components of this discussion, as is the law, as it relates to the specifics of clinical actions.

The questions that capture this area and the dilemmas that accrue to its difficult subject matter include, but are not limited to;

1. Is the public harmed by the mere appearance of impropriety in the relationship that develops between a psychiatrist and ex-patient?
2. Are the rights of patients slighted by admitting that sincere, non-manipulative psychiatrist and ex-patient relationships are possible?
3. Is there specific meaning that can be drawn from this case presentation that relates to benefit or harm to mental health patients?
4. Is it in the general public interest to review the implications of a detailed analysis of psychiatrist-ex-patient relationships?
5. Do the interests of the injured husband preclude or obviate the free will determinations of a wife with a mental health diagnosis, and
6. Do the conclusions of the Medical Disciplinary Board regarding the ethical considerations of mental health patients utilizing a deductive model of medical ethics differ substantially from the conclusions reached with inductive models?

There is no doubt that as the 1980s ended there were numerous articles and renewed interest in the issue of relationships between psychiatrists and patients. The majority of the cases that came to professional attention were situations where the patient would later report harm, or the harm would be reported out by therapists who later took care of the patient. It was acknowledged in the literature of the time that this created real limitations, and that it was entirely possible that patients not harmed were missed in the literature entirely. It was in appreciation of the souring of public opinion that the official position of the American Psychiatric Association took a harder line in seeking to portray relationships following psychiatrist-patient as uniformly derelict.

The positioning of the APA had a very public service to perform, to reassure the public that psychiatry was able to police its own turf. To do so, the public was encouraged to believe that the most enlightened form of medical ethics was conveyed by the position of "Once a patient, always a patient." There could be no termination in the circumstance of a person entering therapy that could ever successfully disentangle the issues of transference. In this very vulnerable population of people, such preventative measures were designed with the protection of the public in mind. Not far from the minds of the policy makers was the disdain of a public informed that their psychiatrists were all too often becoming involved with patients. The aim of the APA, in setting a public tone, was, therefore, twofold. To deliver public safety with an ironclad resolution, and to prepare the public for the idea that psychiatry was committed to do the right thing.

The corollary that expresses the rejoinder implied in the updated APA position might be expressed, as "A patient in one way is a patient in all ways." In reflecting on intrinsic values of a person who has been a patient, the presumption here is there is no provision for the capacity of a person, once a patient, to be possessed of any normal compliment of human emotions that exist separately from the identity of being, first and foremost, a patient.

There was considerable caution about the rapidity with which these medical ethical decisions were recommended, voiced by several past presidents of the APA. But the previous position of the APA was modified away from a caution that relationships in these circumstances were *almost always unethical*, toward a position that held the inevitability of abuse as an accepted fact.

Seemingly, this is a partial answer to our first question, appreciative of the position of the Ethics Committee of the APA. By their reckoning, it would appear psychiatry is wounded by even the appearance of impropriety. Wounded enough to warrant the adoption of a position relatively unique in medical ethics. A deductive model of reasoning that holds the "correct theory" in these circumstances is the unassailable priority of strong paternalism. That only by holding this principle uppermost can the correct interpretation of events occur.

So, we ask again, is the public harmed by the appearance of impropriety in the relationship that develops between a psychiatrist and ex-patient? We think it a more informed answer to conclude that the public is truly harmed when it is encouraged to adopt a stereotypic answer to this question that incorporates a wholesale dismissal of the unique qualifying features of every separate person with a mental health diagnosis that individually determines the correct comprehension of competence, autonomy, quality of life determinations, medical indications and patient preferences, against the backdrop of specifically distinctive contextual features.

To admit that sincere, non-manipulative relationships are possible between psychiatrists and ex-patients does not eliminate the reality of abuse that does occur. Are patients placed at risk if such a moderate position is endorsed?

Our case notes describe a situation where there never was a complaint by the woman involved. The issue of abuse proceeded on the strength of the allegation of abuse, brought by an aggrieved ex-husband, alone. Here, the assumption of theoretical abuse was sufficient to engage legal machinery.

To allow each case to describe its own unique profile, using real experience and engaging an ethical model that allows a synthesis of the simultaneously applicable and competing medical ethics principles, would necessarily mean that (unlike Ms. Clayton) no mental health patient's appropriate ethical rights would be ignored. In a thorough review, there would have been a fair critique of the specific attributes of competence. There would have been an explicit review of the indications, or lack of indications, for a position of strong paternalism. There would have been voice for the concern of patient autonomy. There would have been nothing in this process that disinclined vital autonomy from protection within the principle of beneficence, when it was due. The consciousness that animates an inductive model of ethics is simply less comfortable with a "one size fits all" formula.

The answer to the question is no, patients are not encumbered with additional risk when a more comprehensive model of medical ethics review is employed. A model that seeks to distinguish the nuance of each differing presentation will more accurately protect the ethical interests of those concerned.

There is a significant insight afforded by this case that allows us to see the embedded fingerprint of strong paternalism and its effect on mental health patients. The intensity of emotion that was incurred in the reaction that Averill and Clayton received is evidence of one of the most defining reasons why the occasion of strong paternalism is scrutinized by ethicists, and the occasions in which such a position is sanctioned are so strictly delimited.

In our case presentation, the assumption of wrongdoing was an accelerant to the process of proceeding to conviction (in the Stipulated and Agreed Order), finding Averill guilty of 'abuse of a patient'. This assumption was a natural consequence of the deductive model of ethics which became the applicable ethical standard only after these events, when the APA made the determination that there was one correct, infallible way to interpret all these situations, invariably; the psychiatrist may never become involved with the ex-patient, there may never be a way that a psychiatric patient can be considered an ex-patient.

Logically, that cannot be true unless you are comfortable with the total dismissal of the rights of mental health patients to autonomy, let alone the vital autonomy of spiritually informed decisions. That cannot be true unless you thoroughly dismiss out of hand the possibility that mental health patients possess competence. That cannot stand unless you are willing to dismiss the applicability of quality of life determinations for mental health patients and negate their ability to have a preference. If you state that the whole matter hinges on the doctors' actions, then you are saying that the patients are meaningless.

Which is exactly what happened here. And when the employer, the Board and their investigator pronounced their contempt for the relevance of the patient, according to their deductive model which they were led to believe represented the best that ethics had to offer, *real abuse*, as opposed to *theoretically possible abuse* occurred.

This model of one correct way to adjudicate all ethical questions contains the capacity for reducing mental health patients to irrelevancies. It is not protective. It masquerades as protective. It annihilates a patient's legitimate claim to ethical consideration.

It is in the general public's interest to review the precedents being set here. Do we wish to encourage the type of situation where a woman may be described as being abused by the pronouncements of the State Department of Health and Medical Board, when she knows she has not been so abused? Where, as a consequence of being a mental health patient, when a third party describes her as 'mentally ill,' her comments are dismissed? Where all professional input to counter the assumptions of abuse and incompetence, as prosecutors move to legal advantage, is deemed inadmissible? Where the woman's clearly competent statements are ignored?

The effect of these official actions was to damage the woman directly who the Board loudly proclaimed to the public that they had protected.

Of course, all of this is done in a very public manner under the name "ethics." From the detailing we have outlined in this inductive model of ethical analysis, it becomes clear just how many component pieces of applicable principles were eliminated from consideration. From the detailing of each particular principle that we have illuminated, there is little to recommend what happened to Averill and Clayton as a representative sample of the philosophical rigor and requisite humility of medical ethics.

The injured ex-husband, whose wife determined that the relationship was over, has understandable reason to be upset. He did not see the devolution of the relationship with his wife in the same way that she did. He did not desire an end to the relationship, until the issue of another suitor was raised. His antipathy transferred easily from the unwanted separation, to the person he considered manipulative. He was sure that Averill had stolen his wife, while vulnerable, from him. Believing as he did, that she was acutely mentally ill and unbalanced, his recourse was to seek to destroy the career of Averill.

These sentiments are a normal human reaction. The question arises whether Eason was encouraged in his anger by the invitation to believe in the concrete "infallibility" of the APA's position. The righteousness of his anger certainly found an ally in the deontological connotations of a narrow interpretation of paternalism, a position of the APA that seeks to define away alternative interpretation.

CHAPTER THREE

ONCE A PATIENT, ONCE A PATIENT

Redefining the Issue

Several operative dimensions help to define a particular ethical moment better than one, in isolation. As is usual in the case of an ethical analysis, the issue initially presented for review is not necessarily the final issue that is arrived at, after careful consideration of the attendant and competing ethical principles. The case for thoroughly reviewing an ethical issue of import without bias, the case for utilitarian views, holds this plasticity as a fundamental strength. It is the ability of principle to flex to the exact proportions of the subtleties of the ethical matter at hand. It is not reliant on absolutism and its attendant intolerance.

When this case was initially brought to attention, it was identified as sexual activity with a patient. Clayton then wrote to the board to tell them that she was undoubtedly an ex-patient and that no contact occurred during therapy. The issue was redefined. When the case was moved to a forensic psychologist for review, it was to determine the presence of a mental disease or defect that would impair Averill's ability to practice medicine. The psychologist found no psychiatric diagnosis. The issue was redefined. When the Board proceeded with their process, it became an issue of "Abuse of a patient."

We have completed an inductive review in which the issue is further redefined. We have poured over the facts to determine that Ms. Clayton was not in the midst of a severe psychiatric disorder. We have evidence of her continuous competence from professional observers. We observe that her marriage to Averill is associated in time with the longest period in her life, to that point, of normal mood. Ms. Clayton was conversant with, understood and made decisions about her autonomy from an informed position concerning the competing ethical principles that revolved about her position, defining herself as completing a piece of work in a therapy not designed along, or conducted according to, an analytic school of thought. Averill also discussed the attendant ethical issues with Ms. Clayton at length. Neither party was naive with regard to the decision they took. The requisite threshold for the intercession of strong paternalism, according to standards of medical ethics, was not met. "Vital autonomy" issues, regarding the spiritual, root understanding of Clayton and Averill's relationship as they experienced it, were expediently expunged from the legal review. Nothing in Ms. Clayton's presentation allows the conclusion that she was less deserving to determine that her spiritual understanding informed her decisions regarding Quality of Life.

When we round out our review of the case, it is redefined. Developing the negative and looking at the picture as we have, another issue comes into focus. It does not find abuse between Averill and Clayton. It does illuminate a peculiarly impoverished condensation of ethics currently in vogue in Psychiatry. And it recommends that the perspicacious ethicist would note an area of obvious oversight in the available theory. An ethical psychiatrist would then move to end the era of absolutism in psychiatry, perceiving the capacity of fundamentalism to distort ethics in the service of power, domination, raw emotion and revenge.

Toward a More Balanced Ethical Review Process in Psychiatry

One of the most difficult feats in ethics is retaining balance in the midst of exceedingly emotional real-life scenarios. The emotions that arise naturally from the comprehension that sexual abuse does occur within psychiatrist-patient, doctor-patient, psychologist-patient and therapist-patient relationships has spurred a needed reaction on the part of the respective disciplines to make exceedingly clear the prohibition against this manipulative and exploitative action. Often, in the face of rationalizations on the part of the psychiatrist/doctor/psychologist/therapist, the need for specificity in the relevant law or policy to prohibit such relationships becomes concrete to preclude excuses allowing perpetrators to escape censure. The question becomes whether this move to concretize policy simultaneously removes the ability to differentiate those occasions where genuine attractions have given rise to relationships of substance.

There have been opinions in the literature attempting to inform of the difficulty that arises when policy is encountered by situations where there are legitimate tensions between competing ethical principles. Psychiatrists are not well prepared for these contingencies. It requires an ability to let the certainty of the policy abate while putting in place the corrective lenses of principles, not unlike what we have done in our review of the case preceding. Only by viewing the real-life dilemma with each lens does the picture clarify.

To put the tensions in perspective, if all you had ever been taught about a given ethical situation, the relationship of a psychiatrist with an ex-patient, is to stand fast behind an unequivocal strong paternalism, you would consider yourself a good student of ethics if you responded to a case involving these circumstances as though the only answer conscionable was to unequivocally endorse strong paternalism. To the contrary, ethics is not successfully comprehended by resorting to unidimensional analysis. Every formal student of ethics is instructed otherwise.

With what ethical armamentarium does the psychiatrist equip him/herself when struggling with these questions? What were the sources available to which the psychiatrist at that time could refer? There were a number of resources available at the time of these events to psychiatrists with respect to the opinions of the American Medical Association and American Psychiatric Association. (**See Appendix A.**)

In searching for information, a psychiatrist at the time of these events devoted to examining the possibility of ethical, healthy relationships with ex-patients would be struck by several features in the literature. One, there really didn't exist a definitive statement to the effect that healthy relationships between psychiatrist and ex-patients were categorically impossible. Two, apparently it was conscionable to cautiously entertain being socially related to an ex-patient, in anticipation of doing professional work on a project together. Three, it was possible to become the supervisor of an ex-patient, for the purposes of reviewing their clinical work in mental health, if you had resolved satisfactorily the issues of therapy. And, most importantly, four, there existed no cautionary statement to inform psychiatrists to perspicaciously pay attention to the unique qualities of each situation, encouraging an appropriate trepidation in trying to define difficult ethical issues.

At the time of these events, there was very little help the inquisitive psychiatrist would find in trying to fathom whether the APA had ever considered the merits of a psychiatrist/ex-patient relationship with the multidimensional tools of ethical analysis. It would appear the only explanation that the APA conceived of with regard to these relationships was rooted in the language of exploitation, unresolved countertransference, and with the ascendancy of the term "sexual" as a complete representation of what is possible in human affairs.

Is it logical to describe a marriage as being wholly communicated in scope, breadth and timbre, as "sexual contact with a former patient?" Are all the virtues of partnership, fidelity and trust that are the fabric of a marriage secondary to any compliment that sexuality lends to that marriage? Is a person's sense of spiritual reality cheapened when, in a marriage they perceive as manifesting from transcendent, selfless values, they are informed that any sexual component is deemed improper? Against their explicitly competent, mentally healthy judgment? Clearly there exist no ethical principles to support these suppositions.

If in fact we review the explicit ethical statements that do exist and hold them up to our case presentation, a curious new perspective is advantaged. Dr. Averill did provide care with compassion and respect for human dignity. The truth of this was so fundamentally apparent to Ms. Clayton that she at once distinguished his actions as devoid of manipulation or deceit, and was never intimidated by the vituperative attacks she weathered at the hands of critics. The two, Averill and Clayton, existed in a marriage of equals. There was no sense that one enjoyed a position of unbalanced power over the other. There was no evidence in the period, exceeding a decade, that transference or countertransference were relevant, overarching concepts. Averill did not discriminate, on the basis of Clayton once being a patient, with any prejudice, that this ex-patient was inherently incompetent to know her own heart and soul. As such, he refused to be a party to any policy that excludes, segregates or demeans the integrity a patient. His right to protest social injustice, represented in the wholesale dismissal of patient autonomy, is protected as a valuable ethical contribution, not reflecting negatively on his clinical ethical standards. Further, his decision to be circumspect about the conveyance of confidential information to Ms. Clayton's psychologist, which the psychologist assumed to be primae facie evidence of duplicity, is actively championed in the physician's professional code of ethics, as well as required by HIPPA law. Averill's behavior not to give information without a release is exactly correct, to the letter and spirit of the Annotations, and existing federal Law.

It is possible, although certainly not foreseen at the outset of our review, to gradually come to the conclusion that, although a heretic, Averill behaved in accordance with the ethical guidelines of his profession, holding a minority view that there are exceptions to the useful policies pronounced by the APA for the protection of patients, where patients are best protected by being afforded the right to be considered with dignity, with respect for autonomy, competence, and to be allowed to own wisdom.

The General Assembly of The World Psychiatric Association issued The Declaration of Hawaii in 1977, revised in 1983. The relevant passages from that document that apply to psychiatrist/ex-patient relationships are:

2. Every psychiatrist should offer to the patient the best available therapy to his knowledge and if accepted must treat him or her with the solicitude and respect due to the dignity of all human beings …

5. No procedure shall be performed nor treatment given against or independent of a patient's own will, unless because of mental illness, the patient cannot form a judgment as to what is in his or her best interest and without which treatment serious impairment is likely to occur to the patient or others ...

6. As soon as conditions for compulsory treatment no longer apply, the psychiatrist should release the patient ...

It would be difficult for the inquisitive psychiatrist to get much from this document about the possibility for healthy psychiatrist/ex-patient relationships. It would be hard to mistake the tone of the sentiments, however, that advise respect for the dignity of all human beings, the inadvisability of forcing the issue of treatment and the need to release people from care when they desire. It would be impossible to conclude that these sentiments are an invitation to dismiss the importance of autonomy for mental health patients, impossible to conclude that there would be universal sympathy for the notion "Once a patient, always a patient."

Then there are the offerings of T. Byram Karasu in *Psychiatric Ethics, second edition,* edited by Sidney Bloch and Paul Chodoff, chapter 8, "Ethical Aspects of Psychotherapy". This chapter is 31 pages in length, but again the time spent in developing the area of psychiatrist/ex-patient relationships is scant compared to the depth of the argument appropriately developed against sex with clients.

> "The second misconception is that sexual involvement subsequent to the termination of therapy is neither unethical nor illegal. Quite a number of therapists have used this fact of therapy termination as a defense; but in no instance has a defendant been cleared on this basis ... In any event, the clinician would be on ethically and legally safer ground if he adheres to the motto 'Once a patient, always a patient.'"

Here again the call to strong paternalism is presented as though it is entirely sufficient to canvas and qualitatively replace a duty to beneficence, impugning autonomy as beside the point, creating the image that competence of mental health patients is unrelated to any judgment about their ability to consciously act in their own behalf to terminate therapy and end its jurisdiction in matters of their free will. The literature is deficient. There simply isn't a good discussion of the autonomy of mental health patients to be found. One might logically assume the consideration is irrelevant.

So, it is small wonder that when psychiatrists congregate for the purpose of reviewing a case like our example, they are struggling from the outset to find a balance between the absolutism of policy and an ethical concern for mental health patients, conscious of patient ownership in ethical decision making.

An image only exists when there is contrast that defines its contours. So, what happens to this image of abuse if, in fact, that competent, independent soul describes herself as lucky to have found the person she believes herself destined to partner, spiritually, in this lifetime, in the highest order of honesty that her heart could experience or express? If she knew for a fact, to the depths of her intuition, that she most certainly had not been abused? The answer is that the image of an abuser fades from view, logically, irrevocably.

We set out to involve ourselves in a case study and ethical analysis. And we end up with a discovery of a gaping hole in the firmament of Psychiatric ethics. A hole that is rendered exactly when psychiatrists get so caught up in professional prohibitions that they forget that patients are human beings. And that they, like all other human beings, are capable of finding themselves, suddenly, in the presence of the person they belong with in this lifetime. That moment is felt at a spiritual level of integration that does not ask permission of psychiatry to exist.

Maybe it takes an example as rich in detail as our case presentation to bring this matter into the light. A careful and thoughtful exercise of ethical analysis recommends this area to Psychiatry for reengineering.

The field of Psychiatry is becoming marvelously sophisticated in its bio-logic, neuroanatomic, genetic and psychopharmacologic knowledge base. Increasingly, the faith of the public is being restored to this deserving field of medicine, owing to the industry of the scientists and innovators who are developing these frontiers. The pain and suffering that these efforts alleviate are positive reflections, indicators that psychiatry is increasingly representative of the finest traditions of western medicine.

The sophistication of Psychiatry, however, does not extend to its current forays into the complicated field of medical ethics. It is not enough to cloak those responsibilities with the mantle of paternalism and conclude that the story is complete. If the refinement of psychiatry is to extend to ethics, then the issue of patient autonomy cannot be buried. Put another way, psychiatry may know a great deal about disease states, but their expertise does not routinely extend to the definition of all that is normally possible.

Psychiatry, in its ignoble history, has offered up several notable positions, only to be abandoned when their farcical nature was exposed. Among these: psychoanalytic treatment for Obsessive Compulsive Disorder, ho-mosexuality as a mental disorder, insulin shock therapy for depression, lobotomies. The identification in the early 1990's of as many as 5% of inpatients at a Harvard Hospital as Multiple Personality Disorder, loose criteria for Bipolar Disorder in the 1990's for children that probably mis-identified 20 to 30 times the number of children who would eventually become Bipolar, as suffering from the disorder.

In this current manifestation of malfeasance, psychiatry conflates its authority with established religious tradition, and declares itself better equipped to determine the spiritual correctness of relationships than all the traditions, combined.

Comment: Self-Righteousness, The Conveyance of Intolerance

The face of self-righteousness endlessly morphs. Like an insidious virus, it finds safe haven in the hosts of successive generations. It is difficult to esti-mate the effect this phenomenon has on the evolution of moral and ethical philosophy, but it is safe to say that our review uncovers such an effect in psychiatric ethics. Our search was careful enough to produce the evidence.

Self-righteousness projects an allure that makes it simultaneously popular and rewarding; the image of oneself as absolutely, incontrovertibly correct; the power that this piety invests in action; the certainty of moral sovereignty. The ability to complete a defensible perimeter of truth, through which to perceive the world, and be protected from it. The invitation to define those elements of perception foreign to the perimeter as defective, immoral, baseless. Owning the moral imperative that would reflexively demean the character of anyone with a differing inclination. The encouragement to ridicule dissenters as not possessing similar fine qualities of character embodied by those enlightened with the knowledge of an absolutely correct path.

Somewhere in the emotional circuitry of the brain there must exist a rewarding neurological counterpart of this phenomenon. At a level below and competing with the higher order of abstract reasoning, there must exist a deeply appetitive drive that, once engaged, encourages the dismissal of logical debate. Swooping in to protect its territory, recruiting the kindling of whatever unresolved powerlessness exists in its host to its own devices, this emotional circuitry offers, instead, the promise of conviction and empowerment.

Self-righteousness lent its momentum to the Crusades, encouraged the imprisonment of non-Christians in the Middle Ages. It bore witness to the prosecution of witches at Salem and found expression in the kangaroo courts of vigilante justice. It inhabits the heart of the racist. It abets the world's tensions. It is the fuel of conflict; it is renewed exactly by the ravages of conflict. It is an ancient nemesis of reason. It is enclosed in a cloak of narcissism; impenetrable, regal, splendid.

Like an infection lying dormant at the nerve root, it is encouraged to usher forth under stress. It possesses power sufficient to liquefy the established guidelines of inductive ethical analysis. The dormancy of this pathogen conceals its presence. It is hard to detect. But if you run the assay carefully enough, you will find evidence of this infection of self-righteousness throughout the phrase "Once a patient, always a patient."

For such a phrase may never exist without its attendant underlying assumption, that the doctors' concerns to protect patients are always more important than the patients' right to experience their self-actualization, spiritual centeredness, maturity and intuition as informing them in their autonomous path through their competent lives.

When this is the sum of psychiatry's current proffered contribution to the promotion of ethical standards, who can be blamed for falling into line with an absolutist, deductive model? The investigator at the Department of Health? The advisor who drafted a hardline policy for the Board? The lawyers for the Department of Health, eager for a conviction in a "black and white" case? The aggrieved husband? None of them can be blamed for following what they took as the best approximation of ethics available.

It would be preferable if psychiatry considered the negative impact of hosting this formidable issue. As currently positioned, through the invocation of strong paternalism, the APA actively invites the incarnation of self-righteousness to replace reason in the contemplation of patient autonomy. The emotional fire that is ignited by recourse to such fundamentalist polemics is not so easily extinguished.

Patients are not psychiatry's chattel. Psychiatry, in its sophistication, deserves more enlightened ethical discourse.

APPENDIX

Available at the APA's online web site in the year 2000:

http://www.psych.org/apa_members/ethics_docs.html.
The list includes documents available as of 1999; 19 separate entries were described. Contained within that list are references to: ethics in managed care, a model for ethics curriculum for psychiatric residents, guidelines regarding psychiatrist's signatures, the ethical guidelines of the American Academy of Child and Adolescent Psychiatry, research ethics and human subjects, guidelines for establishing sexual harassment prevention and grievance procedures, the APA fact sheet for memories of sexual abuse, a report of the Task Force on Clinical Assessment in Child Custody, confidentiality and privilege, the code of medical ethics current opinions with annotations of the Council on Ethical and Judicial Affairs of the American Medical Association, procedures governing appeals to the APA Ethics Appeal Board, opinions of the American Academy of Psychiatry and the Law, ethical guidelines for the practice of forensic psychiatry, guidelines on confidentiality, the APA ethics complaint flowchart, information for complainants on the ethics process of the APA, the Opinions of the Ethics Committee on the Principles of Medical Ethics with Annotations Especially Applicable to Psychiatry, 1995 edition, and The Principles of Medical Ethics with annotations Especially Applicable to Psychiatry, 1998 edition.

If you then moved to the web site:

http://www.psych.org/apa_members/ethics_index.html

you would have found access to The Principles of Medical Ethics with Annotations Especially Applicable to Psychiatry. In the forward, all physicians were enjoined to practice in accordance with the medical code of ethics set forth in the Principles of Medical Ethics of the American Medical Association. The forward specifically stated that "special ethical problems" exist in psychiatry, and that the annotations especially applicable to psychiatry "are not designed as absolutes and will be revised from time to time so as to be applicable to current practices and problems."

For our purposes we will review only those areas that directly pertain to psychiatrist/ex-patient relationships, or have specific relation to our case presentation.

Principles of Medical Ethics
American Medical Association

Section 1: A physician shall be dedicated to providing competent medical service with compassion and respect for human dignity.

Section 2: A physician shall deal honestly with patients and colleagues, and strive to expose those physicians deficient in character or competence, or who engage in fraud or deception.

Section 3: A physician shall respect the law and also recognize a responsibility to seek changes in those requirements which are contrary to the best interests of the patient.

Section 4: A physician shall respect the rights of patients, of colleagues, and of other health professionals, and shall safeguard patient confidences within the constraints of the law.

Psychiatry interprets these principles:

Annotations Especially Applicable to Psychiatry

1. The patient may place his/her trust in his/her psychiatrist knowing that the psychiatrist's ethics and professional responsibilities preclude him/her gratifying his/her own needs by exploiting the patient. The psychiatrist shall be ever vigilant about the impact that his/her conduct has upon the boundaries of the doctor/patient relationship, and thus upon the well-being of the patient. These requirements become particularly important because of the essentially private, highly personal, and sometimes intensely emotional nature of the relationship established with the psychiatrist.

2. A psychiatrist should not be a party to any type of policy that excludes, segregates, or demeans the dignity of any patient because of ethnic origin, race, sex, creed, age, socioeconomic status, or sexual orientation.

Re: Section 2;

1. The requirement that the physician conduct himself/herself with propriety in his/her profession and in all the actions of his/her life is especially important in the case of the psychiatrist because the patient tends to model his/her behavior after that of his/her psychiatrist by identification. Further, the necessary intensity of the treatment relationship may tend to activate sexual or other needs and fantasies on the part of both patient and psychiatrist, while weakening the objectivity necessary for control. Additionally, the inherent inequality in the doctor-patient relationship may lead to exploitation of the patient. Sexual activity with a current or former, patient is unethical.

Re: Section 3;

1. However, in other instances, illegal activities such as the right to protest social injustices might not bear on the image of the psychiatrist or the ability of the specific psychiatrist to treat his/her patient ethically and well.

Re: Section 4;

1. Psychiatric records, including even the identification of a person as a patient, must be protected with extreme care. ... Because of the sensitive and private nature of the information with which the psychiatrist deals, he/she must be circumspect in the information that he/she chooses to disclose to others about a patient. ...

2. A psychiatrist may release confidential information only with the authorization of the patient or under proper legal compulsion ...

The 1995 edition of Opinions of The Ethics Committee on The Principles OF Medical Ethics is a 63-page text, published by the APA. It is the re-coding of 161 questions and answers regarding ethical situations. Of these questions and answers, only four relate to the area of psychiatrist/ex-patient relationships:

Section 2-D

Question: Might there be exceptions to the statement in the code of ethics that a sexual relationship with a former patient is unethical?

Answer: *The Principles of Medical Ethics*, adopted by the American Medical Association, and the "Annotations Especially Applicable to Psychiatry," added by the APA, are not laws but standards of conduct, which define the essentials of honorable behavior for the physician.

If a complaint which raises the issue is filed against a member psychiatrist, it becomes the responsibility of the district branch ethics committee to deal with that complaint by careful consideration of all the relevant facts, especially any evidence indicating exploitation of the former patient. The ethics committee will then determine whether the accused psychiatrist has behaved unethically.

Section 2-GG

Question: A psychiatrist was accused by a former patient of sexual misconduct and she was encouraged by her present psychiatrist to file a lawsuit; the psychiatrist also filed a complaint with the licensing board. The original psychiatrist denies his guilt and wants to be heard by his district branch ethics committee who did not receive a complaint. Can he request a hearing?

Answer: There is no provision for a potential accused member to seek a hearing; that action lies with the complainant. Perhaps there should be such a mechanism though the psychiatrist seeking this hearing could be subject to disciplinary actions; he cannot ask for a hearing with immunity.

Section 2-PP

Question: An ex-patient calls a psychiatrist's wife to inform her that she and the psychiatrist are having an affair. Can the spouse bring an ethical complaint?

Answer: Yes.

Section 2-CCC

Question: As a former patient, I would like to have a social, non-sexual, relationship with my former psychiatrist. We are also planning to work on a professional project together. okay?

Answer: Your former psychiatrist should discuss this with a colleague. What is the nature of the social relationship? Will this interfere with getting further treatment if that is necessary? Will there be an uneven control issue on the project that might exploit unresolved issues? Caution is the word to avoid risk for you and your former psychiatrist.

There is only one other comment bearing directly on the case we review, found in Section 2-33, that it may be ethical for a psychiatrist and ex-patient to be related when the psychiatrist becomes a supervisor for the clinical work of an ex-patient who is a therapist, if it can be determined that no transference countertransference issues exist that would harm the ex-patient or lead to misuse of the supervisory role.

ENDNOTES

1   Beauchamp T. L., Childress, J. F. Principles of Biomedical Ethics. New York: Oxford University Press, 1989.

2   Clinical Lecture, August 2, 1993. Seminar in Health Care Ethics, University of Washington, Nancy S. Jecker, Ph. D.

3   Principles of Biomedical Ethics, Tom L. Beauchamp and James F. Childress, 3rd Edition, page 51.

4   Ibid. See Baruch Brody, Life and Death Decision Making (New York; Oxford University Press, 1988), Chapters 1 and 2.

5   Clinical Lecture, Truth -Telling and Promise - Keeping, A.R. Jonsen; Seminar in Health Care Ethics, August 3, 1993.

6   Principles of Biomedical Ethics, Beauchamp & Childress, page 194-195, (Chapter 5 - The Principle of Beneficence).

7   Ibid, page 215

8   Ibid, page 215

9   Ibid, page 210: 21st footnote, page 250

10 Clinical Ethics, 3rd Edition, A.R. Jonsen, et al, page 85

11 Ibid, page 44

12 Principles of Biomedical Ethics, Tom L. Beauchamp and James F. Childress, 3rd Edition, page 82

13 Ibid, page 84

POSTSCRIPT

Day 10,526: Conclusions of Dr. Averill's communication to the Board.

First, thank you for the work you and your colleagues perform in order to ensure that medical providers in the State of (...) are serving the needs of the patients they care for, with all the resultant duties that devolve from that promise. It is no doubt daunting, and I'm sure there is never an adequate instructional manual to allow you to foresee solutions to the unique questions that reach you.

I am writing by way of denouement, neither with malice nor contempt. I believe closure is due, and I think, far from loathsome snark, I can make the process entertaining. Consider the following analogy, borrowed from real life. You may read the analogy straight through first; to be sure I am an apt student of history, not to have fouled those details. But my hope would be to light your interest and willingness to see the parallels to my own history of 25 years ago. The more I thought about this, the more it made sense.

ANALOGY

The House Intelligence Committee has a long and distinguished history of bipartisan cooperation. As such, it is tasked with ensuring Congressional oversight of;

"The United States Intelligence Community (IC), a federation of 16 separate United States government agencies that work separately and together to conduct intelligence activities to support the foreign policy and national security of the United States. Member organizations of the IC include intelligence agencies, military intelligence, and civilian intelligence and analysis offices within federal executive departments."

Currently chaired by Devin Nunes, the Committee received the task of reviewing whether the United States had been attacked by efforts of the Russian government to affect the elections of 2016 in the United States of America, and, in any eventuality, to be a public forum for the examination of this issue, with the clear understanding it had a function to provide answers that would allow Americans to know exactly what happened, what peril we may yet face, and what steps would be necessary to insure the sanctity of upcoming election processes. There was, at the inception of this task, a public display of cooperation between Adam Schiff, ranking Democratic member and former prosecutor, and Mr. Nunes, Republican Chair. There was a seemingly insignificant detail at the time, that Mr. Nunes had been a member of the Trump transition team, an item that was hoped to be inconsequential to the planned proceedings.

As intelligence details surfaced, it became clear that the IC had come to the inevitable conclusion that Russia had been decidedly involved in the election process, and had interceded in a totally provable way with the election, on a number of fronts. Mr. Nunes had a spectacular departure from normal processes of the Committee when, of a Saturday night, he took a cab to the White House, entered a secure intelligence briefing room and received a classified report that names had been unmasked in the intelligence process during intercepts that forced the Attorney General to secure the identities of those involved, and the people so identified were actually involved in the Presidential transition team. Neither Mr. Schiff, nor any other members of the Committee were asked about the propriety of this uniquely awkward overstepping of boundaries, that put Mr. Nunes in possession of material supplied by the White House for the purpose of sowing doubt about the motives of the IC and Attorney General, at a time where the White House itself was quite possibly under, or about to be under investigation. It stands as a totally unique departure from the balance of powers within the government, that a separate but equal Congressional House Intelligence Committee could be so co-opted by Executive Branch motives. And Mr. Nunes ran with this, declaring the matter on National TV from the driveway of the White House. A signature disgrace for the Committee, and an action ridiculed to the degree that Mr. Nunes removed himself, ostensibly, form his Chair position for the purposes of the Russia matter, although that 'recusal' was never something he fully endorsed, and would later be seen to be merely a tactic to remove himself from view to pursue his own motives and a parallel intelligence process among Republican members only. That split constituting serious erosion of any type of bipartisan cooperation to be sure. But by remaining in power behind the scenes, Mr. Nunes had an even more insidious influence in the eventual dissolution of goodwill within the Committee, in as much as Republicans on the Committee would eventually be visible in their sham participation. Mr. Nunes would not allow the normal processes of subpoena for records, emails, accounts, he and Republican colleagues on the Committee allowed all voluntary statements of witnesses to stand without question as satisfactory, far from adequate to the task of getting witnesses to be held accountable for truthful testimony. None of the eventually indicted felons that Special Counsel Robert Mueller had brought to heel in grand jury processes, none of those eventually pleading guilty were remanded to the Committee for testimony, over the objections of the Democrat minority. No bank records were subpoenaed, no due diligence of any sort in the process, the entire investigation brought to a premature closure over objection, and the submission of a public document drafted by Mr. Nunes, which the IC warned against release, as "reckless."

In the Democrat's response, it became clear Mr. Nunes had never made effort to review underlying intelligence documents that laid bare the inadequacy and erroneous conclusions of the eventually public Republican's statement. And that the Republican's statement had been concocted in private, with no input or review from any Democrat on the Committee, secret until put in the President's hands.

The statement's conclusion, that absolved the President of collusion, suited the cherry-picked fiction Mr. Nunes had for so long sought to produce, and the President was able to (ALL CAPS) tweet out the proclamation of innocence he believed he needed/deserved.

END

How am I doing?

Sometimes it is hard to see an embedded figure, until someone points out the trace of an outline.

I am not going to renew my license in (state) ..., nor am I going to involve myself further in becoming emotionally involved in caring about whether you respond to this or not. My character is such that if it were proven to me I was certainly wrong on an issue or opinion; I would move to make it right. First, by admitting responsibility, followed by a sincere apology and genuine dedication not to make that mistake again. I have written some very thoughtful letters to the Board over 25 years, and the only response I ever received, and make no mistake I appreciate the courtesy of that acknowledgement, was to the effect that records are not kept for more than 15 years, and nothing therefore could be done.

(I) will now make a statement, as they are fond of saying in (former British Commonwealth country) ..., 'to your face.' Note to self: In order to leave an abuser, it is necessary to first create a sense of safety. And then place the facts in sight. And then make decisions fully aware that you have no power particularly to affect the abuser, necessarily. Taking care of your own heart and soul, and knowing you behaved with honor and dignity, it will be eventually time to give that abuser no more power over you. When you are still, and able to be in the present, you will be free.

It will always be true that Devin Nunes will be a stain on US history.

It will similarly always be true that (your) actions (26 years ago) to discredit, disparage, dismiss and disallow the thoughtfully competent decision of an intelligent, former patient to exercise her autonomy and enter a healthy marriage, those actions were the proximate cause of a 7 month psychiatric episode for that woman, which no evidence exists would otherwise have occurred.

Tempting to compare scorecards at the end of the round. Seems a doctrinaire, authoritarian approach to an ethical question got you into positions on the course where you repeatedly hit the ball out of bounds. No Marshall in sight, you just continued to spray shots with some abandon. 22 months after a marriage, 11 months after the birth of our first son, your staff wrote a document that is a retreat from the balanced position that existed at the time, to the effect that in order to understand a situation so involved, you would be relieved from the burden to have to review it along standard medical ethical lines of inquiry. Your staff then denied direct contradictory evidence from the woman involved, in effect denying an argument of vital autonomy, that her spiritual understanding of the world, and her perspicacity about the nature of our relationship, was something germane to the discussion. Your move, and it was very out of bounds. Not ethical. Unethical.

Then your staff systematically omitted the expression (shared in direct communication with me) of her long-time psychologist, (a member of the Psychology Board at the time) who said my wife was not abused. That was expunged. Ball out of bounds.

Then you denied the statement of the receiving psychiatrist on transfer, who said explicitly my wife was not abused. That was expunged. Ball out of bounds.

Then a series of errant shots; denying that her older brother had any pertinent perspective when he said categorically you got it wrong, there was certainly no abuse. Or 4 MD's who provided affidavits, or the half dozen senior psychiatric nurses, the equal number of senior psychiatric social workers who said the same thing. Sending me to a Forensic Psychologist who smugly sat for 30 minutes, not recording a word of what I was saying; which at the time was nearly a verbatim transcript of Beauchamp and Childress' Principles of Biomedical Ethics. Not one word recorded. Never presented to your review that there was a totally common-sense way to understand the events that occurred, in the context of medical ethics, and that I was 100% able to articulate that. And had. Never reported to you were the facts that would have distinguished this as something other than the occasion for displacement of disapprobation towards a culprit. Those facts your staff granularly sculpted to allow a prosecution to proceed separate from an accurate picture of what happened.

25, going on 26 years of evidence you were wrong. And in all this time, at every opportunity to respond to thoughtful critique, offered over these years in more than a half dozen missives, never a word from you that you understand the magnitude of your error. Does anyone miss, in current events, just how alluring authoritarianism is to 30% of the public? This issue of mine displays that taste of the narcissism, righteousness and sanctimony inherent in the proponents of authoritarian ideology, and the errors that are intoned by its inception. And it is Exhibit 1 of why modern medical ethics stands in exact countermeasure to deontological ethics and authoritarian ideology. I see the scorecard. There is every reason for you to write a letter of apology to the woman you disserved.

Day Current: No reply.

Henna Mandala Coloring Book